THE METAVERSE
WHAT EVERYONE NEEDS TO KNOW®

THE METAVERSE
WHAT EVERYONE NEEDS TO KNOW®

SCOTT SHACKELFORD
MICHAEL MATTIOLI
JEFFREY PRINCE
AND
JOÃO MARINOTTI

OXFORD
UNIVERSITY PRESS

Oxford University Press is a department of the University of Oxford.
It furthers the University's objective of excellence in research, scholarship,
and education by publishing worldwide. Oxford is a registered trade mark of
Oxford University Press in the UK and in certain other countries.

"What Everyone Needs to Know" is a registered trademark of
Oxford University Press.

Published in the United States of America by Oxford University Press
198 Madison Avenue, New York, NY 10016, United States of America.

© Oxford University Press 2025

All rights reserved. No part of this publication may be reproduced, stored in a retrieval system, transmitted, used for text and data mining, or used for training artificial intelligence, in any form or by any means, without the prior permission in writing of Oxford University Press, or as expressly permitted by law, by license or under terms agreed with the appropriate reprographics rights organization. Inquiries concerning reproduction outside the scope of the above should be sent to the Rights Department, Oxford University Press, at the address above.

You must not circulate this work in any other form
and you must impose this same condition on any acquirer.

CIP data is on file at the Library of Congress

ISBN 9780197759431
ISBN 9780197759448 (pbk.)
DOI: 10.1093/wentk/9780197759431.001.0001

This book is dedicated to our families, with thanks for their support and patience.

CONTENTS

TABLES AND FIGURES	xi
PREFACE	xii
ACKNOWLEDGMENTS	xv
ABBREVIATIONS	xvi

1 Welcome to the Metaverse   1

What is the "Metaverse," and are we already living in it?	1
What are Metaverse enthusiasts saying?	5
What are Metaverse pessimists saying?	6
Separating hype from reality—did the Metaverse bubble already pop?	7

2 Metaverse Economics   10

Does economics even apply to the Metaverse? Does scarcity mean anything there?	10
What is Metaverse supply and demand? Of the Metaverse? Within the Metaverse?	14
Will there be one or many Metaverses?	19

Are we headed toward one Metaverse overlord, a few, or control by the masses? 22

How will the Metaverse and physical world coexist? 26

3 Speech and Expression in the Metaverse 31

What opportunities and challenges does the Metaverse present regarding speech and personal expression? 32

Could the Metaverse face the same content moderation struggles that have plagued social media platforms? 38

Is the legal system ready for speech and expression in the Metaverse? 41

4 Securing the Metaverse 48

What lessons should we learn from cybersecurity failures on the Internet as we look to the Metaverse? 49

If we live more of our lives online in persistent virtual communities, aren't we just making ourselves more likely to be targeted by cyber criminals? 53

Can the Metaverse ever be secured? 55

How do we implement strong security practices? 56

How can we adopt responsible data and privacy protection practices in the Metaverse? 58

How do encryption and VPNs work? 59

How do you maintain adequate protection against cybercrimes in the Metaverse? 59

Are there any unique cyber threats in the Metaverse? 61

In sum, what can I do to protect myself online and in the Metaverse? 62

5 Identity in the Metaverse 66

What does it mean "to be" in the Metaverse? 66

What is "immersion" in the Metaverse? 69
What about "interactivity"? 70
What does "embodiment" in the Metaverse mean? 71
Can I bring all aspects of my personal identity into the Metaverse? 71
Does anonymity exist in the Metaverse? Should it? 74
Will the Metaverse foster and fairly treat diverse participants? 77

6 Privacy in the Metaverse 80

What do we mean by "online privacy" in the Metaverse? 81
Is privacy in the Metaverse any different than other online platforms or digital spaces? 88
Is privacy desirable in the Metaverse? 92
Is privacy even possible in the Metaverse? 97
What will be the role of regulation in protecting privacy in the Metaverse? 99

7 Governing the Metaverse 104

What laws govern the Metaverse? 105
How does Internet governance apply to the Metaverse? 106
How are companies already governing the Metaverse? 108
Are we asking for a tragedy of the Metaverse if we don't get this right?
What lessons can we learn from other digital and real-world contexts? 111
What role should governments have in regulating the Metaverse? 114

8 AI in the Metaverse 119

How are AI and the Metaverse related? 119
Who and what is real in the Metaverse? 121
Does AI undermine reality in the Metaverse? 124
Is AI-generated content real? 129

How will AI bots populate the Metaverse? 129
What rights should AI bots have in the Metaverse? 132

9 Our Meta Future? 136

Inspirations and warnings from sci-fi: Are we hurtling toward Ready Player One, The Matrix, or something better/worse? 136
How can we educate and protect our kids as they grow up in an increasingly Meta world? 138
What comes after the Metaverse? 140
Conclusion 141

NOTES 143
INDEX 172

TABLES AND FIGURES

Table 6.1. Types of Data Exchanged Online   83
Table 6.2. Types of Data That Can Be Collected in the Metaverse   90
Table 6.3. A Data-Sharing Prisoner's Dilemma   96
Table 7.1. Corporate Efforts to Regulate the Metaverse   110

Figure 6.1. Data Flow from an Individual Book Purchase   86
Figure 6.2. Vacuum Products Frequently Bought Together on Amazon   98
Figure 8.1. Evolution of Zuckerberg's Avatar   125
Figure 8.2. Zuckerberg's New Avatar   126

PREFACE

The Metaverse. Depending on how you look at it, it's either the future of the Internet, the next generation of video games, or perhaps "a deeply uncomfortable, worse version of Zoom."[1] In numerous respects, the confusion, exaggerated promises, and hype echo the envisioning of the Internet itself during earlier eras. After all, William Gibson, who coined the term "cyberspace" in his sci-fi short story "Burning Chrome," famously described it, in part, as a "a consensual hallucination experienced daily by billions of legitimate operators in every nation."[2] From this vantage point, the Metaverse may be considered the ultimate realization of Gibson's original idea of cyberspace itself.

According to *Wired*, in many cases, you can interchange "Metaverse" and "cyberspace"—it's not primarily about introducing new fundamental technology but, instead, about innovating new methods to engage with existing technology through virtual and augmented reality. However, these technologies can also facilitate the emergence of new markets and economies, driven by the creation, purchase, and trade of digital assets, such as non-fungible tokens (NFTs). In their ideal manifestation, users could seamlessly carry these assets across platforms, exploring boundless virtual realms with a persistent avatar (potentially featuring personalized creations like your own customized X-Wing from Disney). Although many opportunities already exist for users to, for example, buy virtual goods and play games (like *Fortnite*); that's only one application of the underlying tech.

In short, "[s]aying that Fortnite is 'the metaverse' would be a bit like saying Google is 'the internet.'"[3]

A wide array of firms are pouring investments into Metaverse tech, sparking a digital gold rush, given the significant advantage for early adopters. *Forbes* estimates this opportunity to exceed $1 trillion in potential revenue.[4] These companies include not only Meta (formerly Facebook), but also Nvidia, Unity, Roblox, and even Snap, to name a few. Nike is already developing ways to equip avatars with digital, personalized sneakers. Prices for virtual parcels of land doubled to more than $12,000 in late 2021.[5] There's even now a "United Metaverse Nation," which professes its identity as "The United Metaverse Nation is the first decentralized autonomous organization (DAO) that acts as a virtual country."[6] Yet, since early 2023, there has also been a cooling off of enthusiasm about the Metaverse, mirroring a general wave of pessimism about Web 3.0 products and services including cryptocurrency, NFTs, and the blockchain itself. In other words, the stakes are high, and confusion abounds, particularly in the public, making it—we feel—a perfect addition to the *What Everyone Needs to Know* series.

This book is structured as follows. In Chapter 1, Welcome to the Metaverse, we discuss the definitions and concepts that comprise the Metaverse ecosystem, tracking how the term originated, and how it's distinct from VR to, in general, separate hype from reality. Chapter 2, Metaverse Economics, focuses on the underlying economic issues that both users and firms are navigating in these persistent, digital environments. Chapter 3, Speech and Expression in the Metaverse, tackles thorny issues around speech, content moderation, and expression in the Metaverse, including how various nations are approaching the regulation of this global space. In Chapter 4, Securing the Metaverse, we analyze lessons from past cybersecurity failures and how they may be mitigated in the Metaverse, along with offering practical tips for how to stay safe online in these virtual communities. Chapter 5, Identity in the Metaverse, delves into the protection of personal information and how diversity, equity, and inclusion apply in the Metaverse, among other topics. Chapter 6, Privacy in the Metaverse, focuses on the meaning of privacy in the Metaverse and its parallels with, and divergence from, privacy in existing digital spaces. Chapter 7, Governing the Metaverse, summarizes how laws apply in the Metaverse, along with examining lessons from Internet governance and how companies are creating codes of content that are already regulating these spaces. Chapter 8, AI in the Metaverse, discusses how Artificial Intelligence (AI) and Machine Learning (ML) tools and techniques are being used and misused in the Metaverse. Chapter 9, Our Meta Future? concludes the book by examining technology and regulatory trends to see how the Metaverse may look over the medium and long run, and whether it may be reminiscent of the *Matrix, Ready Player One*, or perhaps something far better—or worse.

ACKNOWLEDGMENTS

This volume is like an iceberg: what you see reflects only a small part of the enormous efforts that lie below the surface. The authors would like to acknowledge the contributions made by the amazing group of graduate students from Indiana University involved in researching and proofing the chapters, including: Yukuah Hao, Sergei, Dmitriachev, Uduak Ekott, James Romano, Drashti Shah, Wenxi Lu, and Anu Thomas. We are also thankful to the staff at Oxford University Press for their support and encouragement throughout the writing process.

ABBREVIATIONS

ADA	Americans with Disabilities Act
AI	Artificial Intelligence
AR	Augmented Reality
ASD	Autism Spectrum Disorder
ATT	App Tracking Transparency
BID	Biometrically Inferred Data
CEO	Chief Executive Officer
ChatGPT	Chat Generative Pre-trained Transformer
CIS	Center for Internet Security
CISA	Cybersecurity and Infrastructure Security Agency
CSM	Continuous Security Monitoring
CVE	Collaborative Virtual Environment
DAO	Decentralized Autonomous Organization
DDoS	Distributed Denial of Service
DLP	Data Loss Prevention
EHR	Electronic Health Record
FTC	Federal Trade Commission
GDP	Gross Domestic Product
GDPR	General Data Protection Regulation
GPU	Graphics Processing Unit
HAL	Heuristically Programmed Algorithmic Computer
IAM	Identify and Access Management
ICANN	Internet Corporation for Assigned Names and Numbers
IoT	Internet of Things
IP	Intellectual Property
IRL	In Real Life
ISO	International Standards Organization
MFA	Multi-Factor Authentication
MIT	Massachusetts Institute of Technology

ML	Machine Learning
MMO	Massively Multiplayer Online
MFA	Multi-Factor Authentication
MR	Mixed Reality
NERC	North American Electric Reliability Council
NIST	National Institute for Standards and Technology
NFT	Non-Fungible Token
OS	Operating System
PII	Personally Identifiable Information
PIN	Personal Identification Number(s)
PTSD	Post-Traumatic Stress Disorder
SEG	Secure Email Gateway
SIEM	Security Information and Event Management
TCP	Transmission Control Protocol
VR	Virtual Reality
VPN	Virtual Private Network
WWW	World Wide Web
XR	Extended Reality

1
WELCOME TO THE METAVERSE

I don't see someone strapping a frigging screen to their face all day and not wanting to ever leave.

~Elon Musk[1]

What is the "Metaverse," and are we already living in it?

In January 2024, the Chinese Ministry of Industry and Information Technology announced the formation of a 60-member working group drawn from industry, academia, and government to define the rules of the so-called "Yuanverse," or a Chinese version of the Metaverse.[2] In so doing, Chinese regulators appeared to be playing a long game, knowing that they could be defining the terms for how over a billion people interact in persistent digital environments for potentially decades to come. But is such attention justified—after all, what is the Metaverse? As was mentioned in the preface, and depending on your perspective, it may be the future of the Internet leading to Web 3.0 (decentralization), Web 4.0 (immersion), or just "a deeply uncomfortable, worse version of Zoom."[3] In many ways, the confusion, overpromising, and hype are reminiscent of how the Internet itself was envisioned in earlier epochs. After all, William Gibson, who coined the term "cyberspace" in his 1980s sci-fi short story "Burning Chrome," famously described it in part as a "a consensual hallucination experienced daily by billions of legitimate operators in every nation."[4]

From this vantage point, the Metaverse may be the ultimate realization of Gibson's original idea of cyberspace itself.

But what *is* it? As *Wired* has argued, in most instances you can use "Metaverse" and "cyberspace" interchangeably—it's not necessarily about a new fundamental technology, but rather new ways of interacting with existing tech through virtual and augmented reality. Decades after Gibson's *Neuromancer*, there appears to be something of a consensus on the definition of cyberspace as "the complex environment resulting from the interaction of people, software and services on the Internet by means of technology devices and networks connected to it, which does not exist in any physical form."[5] The term "metaverse" was introduced by another sci-fi author, Neal Stephenson, in his 1992 novel *Snow Crash*.[6] An exact definition was not included, but Stephenson likened the idea to "a persistent virtual world that reached, interacted with, and affected nearly every part of human existence."[7] Among the greatest punishments in such a world was banishment from the global network, the closest analogue to "death" that the Metaverse could offer.[8] The dystopian, all-encompassing Metaverse in *Snow Crash* made life for the millions regularly plugging into the network that much worse than the "real" world that featured an apocalyptic pandemic.

So, is the Metaverse just another term for cyberspace? Or perhaps it is best characterized as a specific version or advancement of the concept, namely an immersive cyberspace. Expanding on this notion, one might think the Metaverse to be "virtual-physical blended space in which multiple users can concurrently interact with a unified computer-generated environment and other users, [a vision] which can be regarded as the next significant milestone of the current cyberspace."[9] Since *Snow Crash*, there have been a variety of other popular culture creations that may fit such a notion of the Metaverse, particularly *The Matrix* and *Ready Player One*, discussed further in Chapter 9.

Given this context, we offer the following definition for a metaverse, allowing for the possibility of more than one such environment. A *metaverse* is a persistent, immersive, and widely utilized, virtual world. Our definition has several components, so we complete it by unpacking each element, starting with the notion of a virtual world. On its own, a virtual world is a computer-based, simulated environment through which users can interact with one another. This alone is not enough to be a metaverse—that is, a virtual world encountered by five friends at home using computers, mice, and keyboards does not suffice. To understand why, consider the label of "widely utilized." A metaverse must have many users engaging it in some way, which could include social interactions, commerce, and education. Note that we stop short of using the descriptor ubiquitous, meaning we believe a virtual world can be a metaverse even if not *all* possible participants partake.

We now turn to "immersive," by which we mean deep absorption—in this case, deep absorption in a virtual world. It is here where virtual reality (VR) and augmented reality (AR) can play an important part in a metaverse as seen with Apple's approach to what they term "spatial computing" with its Vision Pro headset; however, immersion need not only come from such technologies. Finally, consider the last element, "persistent." For a virtual world to be a metaverse, it must have a reliable continuation of existence, as opposed to fleetingly emerging and disappearing, perhaps with the flip of a switch. From here on, we'll generally reference "the Metaverse," with the understanding that the continuing coexistence of multiple metaverses is a possibility, even a probability.

With the Metaverse still in its infancy, it should be noted that cyberspace already touches on different aspects of the "real" world, such as through the growing prevalence of Internet of Things (IoT) products and services, from smart watches and cars to interconnected air traffic control systems.[10] Cyber attacks on these services can have—and unfortunately already have had—real-world consequences. The Metaverse, though, can take this blending of the online and real worlds to a new level, multiplying opportunities, and, potentially, vulnerabilities to be exploited for commercial or geopolitical gain.

A diverse range of firms are investing in Metaverse tech, causing a digital gold rush since there is such a substantial first-mover advantage to be won. *Forbes* has estimated this to be north of a $1 trillion revenue opportunity.[11] The firms include not only Meta (formerly Facebook), but also Nvidia, Unity, Roblox, and even Snap, to name a few. Companies like Nike are already developing ways to equip avatars with digital, personalized sneakers. Prices for virtual parcels of land doubled to more than $12,000 in late 2021.[12] There's even now a "United Metaverse Nation," which professes its identity as follows: "The United Metaverse Nation is the first decentralized autonomous organization (DAO) that acts as a virtual country."[13] In other words, the stakes are high, but the path forward is highly uncertain.

The situation is only getting more complicated with the crypto crash of 2022 and the concurrent decreases in NFT prices and slowdown in Metaverse initiatives more broadly.[14] The acceleration away from Metaverse investment continued through 2023 with the announcement by Disney to cut $5 billion in costs through 7,000 layoffs, including the elimination of its Metaverse unit.[15] But, despite these bumps in the digital road, are we already living in the Metaverse? If not, will we not be until everyone on the planet is wearing the equivalent of Google Glass? That seems like an unrealistic metric, given that no one disputes the arrival of the Digital Age, even though more than three-billion people are still not online as of this writing.[16] By some estimates, more than 400 million people are using versions of the Metaverse as of 2023,[17] though there are few reliable statistics in this regard. The growing use of AR and VR to access a range of Metaverse applications and platforms seems to be a growing trend, despite the "Metaverse winter" beginning in 2022.[18] After all, Apple's CEO, Tim Cook, is going "all in" on its 2024 Vision Pro, with some early adopters like the movie director James Cameron describing his experience with the product as "religious."[19] The reelection of Donad Trump also caused a surge of interest in cryptocurrency and Web 3.0 businesses with the expectation that the industry would be embraced by the new Administration. Yet it is an open question whether the concept itself will become an important driver of commerce and innovation beyond the associated Metaverse hardware and software, which in time could take on a life of its own, shaping social, economic, and political trends. The power of prediction in this regard often fails even proven thought leaders such as Paul Krugman, who famously—even if half facetiously—stated in 1998, "By 2005 or so, it will become clear that the Internet's impact on the economy has been no greater than the fax machine's."[20]

What are Metaverse enthusiasts saying?

There has been no shortage of hype and enthusiasm about the promise of the Metaverse. Without doubt, the most prominent display of optimism was Mark Zuckerberg's decision to change the name of his (at the time) $300+ billion company from Facebook to Meta. Such a bet makes sense after hearing him express his optimistic view of the Metaverse as follows:

> We hope to basically get to around a billion people in the metaverse doing hundreds of dollars of commerce, each buying digital goods, digital content, different things to express themselves, so whether that's clothing for their avatar or different digital goods for their virtual home or things to decorate their virtual conference room, utilities to be able to be more productive in virtual and augmented reality and across the metaverse overall.[21]

Mr. Zuckerberg is by no means alone in the optimism he has expressed and demonstrated. In responding to skepticism about the future of the Metaverse, Richard Entrup, Head of Emerging Solutions at KPMG, commented,

> Having personally witnessed the transition from a no Internet world to a globally connected Internet world, I find it funny to hear the same negative comments being made about the metaverse. I recently gave a talk on the subject. Many older IT executives expressed concern that our children are already on social media and gaming too much and that the metaverse will only exacerbate this issue. I'm not interested in exploring why it won't work or why we shouldn't do it. Web3 is a boat I don't want to miss and one I believe large companies shouldn't miss either. I believe we can achieve revenue growth,

competitive advantage, and improved customer experience through the adoption of the metaverse.[22]

Still others, such as David Clark from MIT's Computer Science and Artificial Intelligence Laboratory, acknowledge the uncertainty about what is to come with the Metaverse: "Where will the standards come from? How open will the system be? Who controls it?" However, Clark holds no doubts about the emergence of the Metaverse over the next decade and a half: "My uncertainty about the metaverse is not whether we will have 'something' by 2040, but what character it will have."[23]

We conclude this amalgam of Metaverse enthusiasm with the expressed expectations of Tim Sweeney, CEO of Epic Games. His vision is among the most expansive—"The Metaverse is going to be far more pervasive and powerful than anything else"—but also comes with a warning: "If one central company gains control of the Metaverse, it will become more powerful than any government and be a god on Earth."[24]

What are Metaverse pessimists saying?

Although the signs were there, beginning in 2022 there was a growing chorus of voices lamenting either the end of the Metaverse, or at least questioning if the hype would ever match reality. For example, Meta laid off more than 11,000 employees that year, and comparisons with massive multiplayer games are useful only to a point. As one example, "Decentraland self-reports no more than 10,000 active users at its daily peak, and that's compared to the nearly 60 million daily active users playing Roblox."[25] In short, despite the optimistic take of entrepreneurs and venture capitalists who have a vested interest in the outcome of the technology, many Metaverse platforms remain immature novelties, and niche ones at that.[26]

At its core, the bigger question about the utility of the Metaverse still looms—will people prefer to interact in persistent, widely utilized, digital environments instead of the real world? Do users crave such a level of immersion, or will it just exacerbate the social and mental health costs that have already become all too apparent in the social media context? Many are unconvinced, such as Strauss Zelnick, CEO and managing partner of the private equity firm ZMC, who said, "I'm skeptical that we're going to wake up in the morning and intentionally sit at home, strap on our headsets, and conduct all of our daily activities that way. We had to do that during the pandemic, and we don't really like it so much."[27] Further, even if the optimists are correct, would the triumph of the Metaverse wind up being a net positive for society? The US National Institutes of Health, for example, has already warned that the Metaverse "can harm children's mental health, such as exacerbating depression, anxiety, addiction, self-harm, suicidality, or anorexia."[28] In addition, there are risks of cyberbullying and sexual exploitation, along with gambling and security issues, that could be particularly hurtful to teenagers.[29] Aside from health concerns, there are also the real risks that failures and gaps in AI governance could be especially problematic in the Metaverse, with Murat Durmus, CEO and founder at AISOMA AG, saying, "The Metaverse is the ideal playground in which the AI can let off steam. The more humans lose themselves in it, the more the AI will take control. That much is certain."[30]

Separating hype from reality—did the Metaverse bubble already pop?

As previously mentioned, starting in 2023, there was a growing consensus that the Metaverse bubble had indeed popped, with many commentators writing its epitaphs. For example, one obituary marked its demise less than three years after its reported birth due to "a lack of coherent vision," which reached a crescendo in the rush to generative AI.[31] A prime example of this trend is when Meta itself walked away from its "Metaverse first" mission. So, what happened? What can we learn from other tech bubbles? Have the reports of the death of the Metaverse been exaggerated? And what future is there for a persistent, immersive, and widely utilized, virtual world?

In the heady days of October 2021, when Zuckerberg announced that the Metaverse would be the future of the Facebook ecosystem, renamed his company "Meta," and promised an immersive version of the Internet in which it would "feel like you're right in the room" with other users, the future of the Metaverse seemed bright.[32] But, despite glowing press at the time, including a 5,000-word feature in *Verge*, it didn't take long for the hype to fade and the problems to mount, among them buggy tech and a vexing identity crisis.

First, consider the technology. Accessing the Metaverse still requires a VR or AR headset, which, let's face it, "is costly, clunky, hard to use and makes you look like a visitor from outer space."[33] Despite selling some 20 million Quest headsets by Spring 2023, there simply were not enough active users to maintain demand for services leading to plummeting virtual real estate prices.[34] Plus, the available experiences, such as Meta's "Horizon Worlds" platform, were so buggy that even its employees stopped using it.[35] But, as discussed below, as new platforms are rolled out and processing power improves, this hurdle will likely be overcome. Consider, for example, the late-2023 rollout of photorealistic avatars[36] and the 2024 introduction of Apple's Vision Pro spatial computing headset.

Second, there is the business case. Despite the calls for the Metaverse to become the "successor to the mobile Internet," precious little attention was given to developing "a clear use case, a target audience, and the willingness of customers to adopt the [Metaverse] product."[37] After all, virtual worlds are not particularly new. There was a variety of 1990s-era massively multiplayer online role-playing games, such as EverQuest.[38] What seemed new in this case was the interface, but that did not slow the initial virtual land rush mentioned above with major companies from Walmart to Disney and Microsoft investing in the technology. Gartner predicted that, by 2026, 25% of Internet users would spend at least an hour per day in the Metaverse, and McKinsey forecast that the new platform would generate up to "$5 trillion in value."[39]

By 2023, the veneer was coming off the original Metaverse bubble with some services such as Decentraland claiming a mere 38 daily users in its $1.3 billion, crypto-based ecosystem.[40] But the real death knells in this first Metaverse phase were a slowing economy and the rush to generative AI platforms as "the next big thing" in tech. As a case in point, Zuckerberg declared in March 2023 that Meta's "single largest investment is advancing AI and building it into every one of our products."[41]

So, was the Metaverse craze just the effort of a single CEO to drive up his company's share price with a half-baked vision and clunky tech? As we discuss later in this book, the lessons from Meta's failures are important to examine and put into the context of other tech bubbles that have come before or are on the horizon (a topic explored further in Chapter 9). But the idea of persistent, immersive, and widely utilized virtual worlds is not going away. Indeed, there is growing excitement about Apple's new Vision Pro headset, featuring spatial computing and a "reality dial," and the buy-now-pay-later options that will go along with it, in helping to define a still-nascent market.[42] It's possible that Apple may help usher in the next stage of the Metaverse that Meta itself has not quite yet been able to realize—if it can solve the technical and business case questions that have bedeviled other firms.[43]

2
METAVERSE ECONOMICS

Buy land, they're not making it anymore.

~*Mark Twain*

There is no shortage of predictions about the economic impact of the Metaverse, both optimistic[1] and pessimistic, as was mentioned in Chapter 1.[2] Any such forecast largely rests on assumptions about the underlying economic mechanisms of the Metaverse. In this chapter, we tackle questions surrounding the economic framework behind the creation of the Metaverse, the economics of consumption and value creation within the Metaverse, the factors behind the number of metaverses and who controls them, and the potential substitution between real-world consumption and digital consumption in the Metaverse. Our aim is not to make a grand prediction about the change to gross domestic product (GDP) spurred by the Metaverse, but rather to highlight the relevant economic theory and apply it to key early facts about it.

Does economics even apply to the Metaverse? Does scarcity mean anything there?

Economics is the science of making decisions in the presence of scarce resources.[3] Scarcity means that human wants for resources (goods, services, and natural resources) exceed what is available.[4] Fleshing out the idea of scarcity a bit further, at its core, it means that, if every good, service, and resource were free, the amount people would want (amount demanded) would exceed the amount that was provided (amount supplied). Virtually any "good" you can imagine fits this bill. As a simple example, consider cars. If cars were free, how many would people want to have, and how many would people produce? As the former certainly exceeds the latter, we have a scarcity of cars.

Price typically plays a central role in resolving this discrepancy between demand and supply. When consumers and producers are allowed to freely trade, price incentivizes production (higher prices make it more attractive to be a producer) and rations the amount produced (only those willing to pay the price receive the product/service). The "invisible hand" pushes the price up from zero to a price point where the amount of a product that people demand at that price is equal to the amount producers supply, known as equilibrium.

The idea of scarcity is closely tied to costs. If a product or service that consumers value requires some cost to provide, there will be little incentive to produce it for free, resulting in scarcity. Even if you think some things will be produced for free due, say, to altruism, in the physical world the costs eventually become prohibitive. For example, you may choose to grow vegetables in your backyard garden and give them away for free, but the amount you can give away is limited by the number of vegetables that it is physically possible to produce in your garden. For these reasons, it typically will take a positive price to resolve scarcity, that is, to reduce the amount demanded (from the amount demanded were the goods free) to a level that producers are willing and able to make.

The digitization of many facets of the marketplace has added an interesting wrinkle to the notion of scarcity. For example, consider the popular game Candy Crush, played every day by millions of people (including the likes of Treasury Secretary Janet Yellen) for fun, to relieve stress, and/or to avoid their kids. Each round of the game requires use of a "life." If a player runs out of lives and wants another, what is the cost of the developer to give that player one more life, that is, one more chance to crush some candy on, say, their smartphone? It's fair to say that it is vanishingly small—effectively zero. Hence, contrary to what we generally would find in the physical world, we have a situation where it is plausible that a producer would be willing and able to produce something (an additional Candy Crush life) for free.

So, is there no scarcity when it comes to digital goods and services? Some still have physical constraints. For example, digital services that require a person's time face constraints; a telemedicine doctor can only conduct so many half-hour consultations in a day. Some digital products have costs associated with the use of intellectual property. When you listen to a copyrighted song on a subscription music service, the service provider generally must pay fees to various parties—including the song writer and performer—for the right to play you the song.

Nonetheless, under some very specific circumstances, there could be a lack of scarcity for some digital goods and services. Consider a digital world with multiple firms producing (roughly) the same digital product where each new copy is effectively costless. Of course, such digital products (like Candy Crush lives) don't appear from nowhere; developers must incur costs to design and construct the game, not to mention hosting it in power-hungry data centers. In some cases, those costs are entirely sunk, meaning if the developers decide to quit operating the game entirely, they get none of the costs incurred to build the game back. With that backdrop, suppose we have a scenario where multiple firms offer the same online game, where all their development costs are sunk and the cost of selling another copy is effectively zero. In such a situation, it is possible that the developers would be willing and able to supply any number of copies of the game for free, meaning whatever number of copies consumers would demand at a zero price would be satisfied—that is, no scarcity!

However, the rosy conditions we just described that can lead to no scarcity are rarely met. Often, firms invest in research and development that is difficult or impossible (due to intellectual property protections, like patents and copyright) to replicate by other firms. With limited competition, it is in firms' interest to produce less than the amount that would satisfy everyone at a zero price. The reason for this constrained production is clear—with fewer units available than what people would want at a zero price, the price would go up to ration the available units, and the firm consequently makes money.

Non-fungible tokens (NFTs) are a recent development that prevents replication. In short, NFTs are a means of making a digital product unique (via a unique identifier) and authenticating asset ownership. "NFTs are valuable precisely because they create artificial scarcity around things that are for the most part identical to works digitization has made available to the masses for free."[5] A caveat to NFT-generated scarcity is that consumers may not care about unique identifiers and simply treat all digital products that are identical but for their unique identifier as being fully interchangeable; in such cases, NFTs per se would not lead to scarcity of digital products.[6]

Even if a firm faces stiff competition for its digital product or service, if it is also able to recover some of its investment, it may choose to leave the market altogether rather than stick around and produce for free. For example, app developers may purchase significant amounts of hardware (computer equipment and infrastructure) to run and maintain their product(s); the costs of these investments generally are at least partially recoverable through resale. When such costs are recoverable, firms will choose to exit and recover their investments until the competition among the remaining firms has decreased to the point where they can raise prices enough to cover those costs. Scarcity again!

The Metaverse has both physical and digital components. Several physical building blocks have emerged over time, most notably virtual reality (VR) and augmented reality (AR) headsets as discussed in Chapter 1.[7] VR headsets allow users to fully immerse themselves in a digital world, whereas AR headsets overlay digital landscapes on top of what you see in the real world. These technological advancements are expected to operate as the "portals" to the Metaverse, in the same way a computer or smartphone is a "portal" to the Internet. Essentially everything *within* the Metaverse is digital, including the virtual worlds that comprise it (their dimensions and features) and the products and services exchanged and consumed within it. Within the digital universe created by the Metaverse, consumers can purchase gaming subscriptions or upgrades, apps, digital property, and more.

For the physical components of the Metaverse, the well-known idea of scarcity in the physical world applies. The construction of these physical components requires, at the very least, physical materials which entail some cost to acquire. In contrast, just like pre-Metaverse digital products and services, goods and services within the Metaverse may defy scarcity under certain conditions. However, again there is reason to believe such conditions won't be met in practice, as firms are investing, and will invest, in digital products and services within the Metaverse that are difficult or impossible to replicate. NFTs are just one example. Another familiar example from the physical world is an exclusive club membership. Here, firms make investments (e.g., influencer endorsements, exclusive offerings to members, etc.) that build the "brand" of their club and are difficult for other firms to mimic. In addition, they commit to exclusivity by capping the number of members; this, in and of itself, can create value in the memberships. In such a case, even with all member activities happening digitally, the difficult-to-replicate investments and the inherent value of exclusivity create conditions where firms will have an incentive to make their digital offerings scarce even if the cost of adding another member is virtually zero.

What is Metaverse supply and demand? Of the Metaverse? Within the Metaverse?

At a basic level, the supply and demand of the Metaverse has two elements. There is supply and demand *of* the Metaverse—effectively Metaverse infrastructure. And there is supply and demand *within* the Metaverse—the goods and services that are transacted by entities inside the Metaverse virtual worlds.

Metaverse infrastructure has both physical and digital components. Taken together, these are comprised of the hardware needed to access the Metaverse (e.g., VR/AR headsets), the software used to generate the virtual worlds, the data storage and network infrastructures, and blockchain and cryptocurrency technology, among others.[8] The suppliers of the Metaverse can fall along several levels of the overall supply chain, and some major tech giants (such as Meta and Apple) have already made investments in multiple levels of this supply chain.[9]

There are three main physical components associated with the supply of the Metaverse: consumer devices, compute, and network capacity.[10] To elaborate:

- *Consumer devices* include VR/AR headsets, phones, sensors, etc. As of this writing, this component is experiencing "rapid technological progress due to learning effects and related supply-side economies of scale."[11]
- *Compute* consists of the server infrastructure required to enable the "enormous amount of information processing required to simulate virtual environment." It usually involves both a fixed cost (costs firms incur no matter how much they produce) as well as variable costs (costs that vary with the level of output), that scale with the amount of activity.[12]
- *Network capacity* refers to the real-time data transmission capacity. Like compute, this component also requires a fixed cost and variable costs that scale with the Metaverse activity.

Digital supply of Metaverse infrastructure largely consists of virtual world-building software and blockchain technologies. Gaming companies like Epic, Roblox, and Unity were early developers of Metaverse-related technology, as many of their platforms allow users to participate in simultaneous, persistent gaming experiences.[13] For example, Roblox users have the opportunity to generate their own "worlds" and build their own Metaverse experiences, generating opportunities for individual wealth creation and innovation within the platform itself.[14] Meta and Apple are also attempting to build Metaverse software in addition to their Metaverse hardware development.

An important element of Metaverse infrastructure supply will be the complementarity between Metaverse hardware and software. This notion is as old as computing itself: computing hardware are just idle materials without software that works on it, and software cannot be utilized without supporting hardware. After all, what good is a personal computer with no operating system (e.g., Windows, OS X) and software applications like MS Word or Adobe Acrobat; and what good are those operating systems and software applications without a computer to run on? Suppliers of computing products have taken different approaches to hardware/software complementarities, with Apple choosing to directly cater to them by producing both in house; and other firms, like Dell and Microsoft, choosing to focus just on hardware or software, respectively. How firms view the hardware/software complementarity for the Metaverse will almost certainly influence their supply chain decisions and ultimately the landscape of Metaverse suppliers (more on this later in the chapter).

Supply within the Metaverse is digital and consists of the products and services exchanged and consumed within the persistent, digital environment. For starters, this would include e-commerce. That is, products people currently buy online (e.g., shoes, books, etc.) can, and almost certainly will, be offered through the Metaverse. However, e-commerce vendors can offer new services in the Metaverse that will enhance the shopping experience; perhaps most notable are services that use the immersive Metaverse technologies to substitute for physical information gathering. For example, consumers will be able to virtually tour homes, assess the fit of new furniture in their house, and try on clothes (among other things) when considering these new product purchases.

Additional possibilities for products and services within the Metaverse are endless, but there are a few categories (beyond information gathering) that immediately stand out. The most obvious is gaming, as this is already in progress. With software infrastructure in place, firms will be able to offer immersive games that can be played within virtual worlds. Another category is virtual human services, including virtual consultations and virtual meetings. These services will facilitate human interaction by eliminating the need for the participants to travel to the same location. They can also provide helpful virtual surroundings and complementary services (e.g., a tranquil virtual setting for a therapy session or 3-D representation of health data).[15] We conclude by highlighting a category that we call Metaverse-specific experiences. These include products and services offered within, and made possible by, the Metaverse. Examples include virtual travel and virtual social environments (e.g., a virtual picnic on "the moon"), standup comedy acts, and concerts.

Demand in the Metaverse is generated by the value that individuals, businesses, and investors place on Metaverse technology.[16] While it can be a major challenge to fully measure how much consumers value the Metaverse itself, a major factor in any consumer valuation will be network effects, both direct and indirect. Direct network effects exist when an individual's value of a product changes depending on the number of other individuals that own that product. A simple example where direct network effects operate is the Internet itself. The value you can get from the Internet generally is greater when more people are using the Internet; as just one example, it means more people you can easily communicate with using email. Indirect network effects exist for a product or service when increased ownership leads to more complementary products and services, leading to a higher consumer valuation. A simple example of indirect network effects involves gaming consoles. As more consumers choose to buy a particular gaming console (e.g., PS5), game developers have an incentive to produce more games (complementary products) for that console; and with more available games for it, consumers will find that console more valuable.[17]

Metaverse infrastructure will have both direct and indirect network effects like those we've seen for the Internet and gaming consoles. These exist both for Metaverse hardware and software. Consider first direct network effects. As more people own, for example, Metaverse VR headsets and have subscriptions to Metaverse software, you will have more opportunities to engage with others in the Metaverse if you own a headset and subscribe to the software. Hence, just like email (and even the telephone before it), you will likely get more value from purchasing Metaverse infrastructure products and services when others are doing the same. Regarding indirect network effects, greater ownership of Metaverse VR headsets and higher numbers of Metaverse subscriptions will create greater incentives for developers to create complementary, Metaverse-facing products and services like those we described above. With more products and services available within the Metaverse, you again will likely get more value from Metaverse infrastructure purchases.

Metaverse precursors in the form of virtual-world gaming platforms have shown promise in terms of consumer demand. In the first quarter of 2021, consumers spent nearly 10 billion hours on the gaming platform Roblox, and more than 42 million users logged in each day.[18] Metaverse-like gaming experiences, such as Roblox and Fortnite, have generated millions of dollars from within-platform purchases and are early adopters of virtual social experiences, like concerts or art installations.[19] For example, in some of the first virtual events of their kind, Gucci launched a two-week art installation on Roblox that attracted 20 million visitors, and a Travis Scott concert held within the Fortnite platform drew more than 12 million concurrent views.[20]

Despite this heightened demand for virtual-world experiences on popular platforms, especially during the COVID-19 pandemic, only 1% of consumers surveyed indicated that they preferred to attend concerts, trade shows, or learning events in the virtual world as opposed to the physical world. However, almost 60% of consumers indicated that they prefer the experience of at least one activity in a virtual world more than the physical world, with most consumers ranking "connecting with others" at the top of the list of activities they are excited about experiencing in a virtually immersive world.[21]

Hence, there appears to be a baseline level of Metaverse demand. For this demand to really take off to where the Metaverse has massive participation (akin to the larger Internet), it will depend on the interplay between infrastructure enhancements (e.g., improved VR/AR headsets) and the activation of both types of network effects—all of which will lead to higher valuation of Metaverse technologies.

Will there be one or many Metaverses?

At its height in late 2022, there were reported to be more than 160 firms that were creating different versions of the Metaverse.[22] By 2023, although there has been a proliferation of "Metaverse-style" platforms, a dozen had risen to some prominence, including Altspace VR, BlueJeans, Cryptovoxels, Gather, Metahero, Meta Horizon Worlds, Nvidia Omniverse, Roblox, Rooom, Sandbox, Second Life, and Somnium Space.[23] These platforms run the gamut in terms of their scope and functionality, which ranges from Rooom's virtual showrooms to Cryptovoxel's virtual world gaming powered by the Ethereum blockchain.[24] It is a common practice for users to enjoy different Metaverses for different needs, including work and play.

Recall from Chapter 1 that our definition of a Metaverse is a persistent, immersive, and widely utilized, virtual world. This definition does not preclude the possibility that there will be multiple metaverses, as opposed to a single, "Metaverse." Whether we ultimately have multiple metaverses or just one will depend on several factors; here we highlight two categories—technology and demand-related. These categories, and the specific factors we consider within them, are not without precedent. Similar factors have been and continue to be at play for a variety of other technologies, as we will make clear throughout this answer.

We begin with technological factors, starting with the interoperability of different virtual worlds. Put simply, the extent to which there are multiple metaverses vs. a single Metaverse will depend on the technological feasibility and cost effectiveness of separate, but interoperable, virtual worlds. Greater ease of interoperability should push us toward a single Metaverse; however, as we'll discuss further in the next section, the converse need not hold. That is, the emergence of a single Metaverse need not imply easy interoperability.

There are three main areas of technological challenges to interoperability: technology inconsistencies, data synchronization challenges, and security concerns. *Technology inconsistencies* stem from the diversity of technologies, programming languages, and data formats utilized by different organizations and systems. For example, interoperability for electronic health records (EHRs) is pivotal for optimizing healthcare quality and efficiency. When hospitals and clinics use different healthcare interface standards, it creates challenges for physicians to gain comprehensive insight into patients' medical histories from external organizations.[25] Other analogous scenarios include the compatibility issues between applications running on Windows versus Linux. With multiple firms working on Metaverse operating systems as of this writing, similar challenges appear likely for the Metaverse.[26]

Data synchronization challenges pertain to the complexities of harmonizing data across multiple systems. Given the massive amounts of data involved with any Metaverse engagement, data synchronization is crucial because it ensures "accurate, secure, compliant data, [and] congruence between each source of data and its different endpoints."[27] Lastly, *security concerns* arise when sharing data between systems without proper security management. Risks can arise in many ways; for instance, when multiple systems are interconnected for interoperability, it expands the potential attack entry points, increasing vulnerability to exploitation. Additionally, the broader access granted to authorized users and systems introduces authentication risks.

Economies of scale can also point toward a single Metaverse. These exist when the per unit cost of production declines as a firm produces more. A classic scenario where this happens is when production has high fixed costs (recall these are costs firms incur no matter how much they produce) and low marginal costs (the cost of producing an additional unit). The Metaverse is likely to exhibit economies of scale, with large fixed costs of development for both the hardware and software and then relatively low costs of producing an extra headset, and particularly for adding another unit of software for consumer access and use. At the extreme, this means it will be more cost efficient to have a single Metaverse as opposed to several or many.

Another key technological factor is the costs associated with Metaverse hardware and software, which ultimately impact the incidence of multihoming. In the Metaverse context, multihoming occurs when consumers engage with more than one metaverse. An analogy is multihoming in gaming consoles, where a consumer multihomes if (s)he owns and plays with a PS5 and an X Box. If multihoming is easier for consumers, this would increase the likelihood of multiple metaverses.

A fundamental demand-related factor that determines whether consumers multihome is the price they must pay to do so. Going back to the gaming example, if a PS5 and X Box both sold for $10,000, we'd expect substantially less multihoming than if they sell for $300. As a general matter, the cost of production will impact the market price—higher costs tend to result in higher prices. Hence, if Metaverse technology proves very costly to produce, market prices likely will be higher, making multihoming more difficult. Thus far, device pricing has varied considerably, with Meta Quest VR headsets priced in the $250 to $600 range in early 2024, while Apple's Vision Pro was released with a $3,500 price tag.

Whether consumers will multihome depends not just on the price, but also on their preferences. Multihoming can come with other costs to consumers, including the time and effort to learn two different operating systems and physical devices, the storage costs of keeping two sets of equipment, and the time costs that come from switching from engaging in one metaverse to another. Hence, even if Metaverse prices are relatively low, this is not a guarantee that consumers will readily choose to multihome across multiple metaverses.

Another demand-related factor that will likely influence whether there are one or many metaverses is network effects. Recall that Metaverse technology will almost certainly exhibit both direct and indirect network effects, meaning that one's value of it increases as others buy and use it. Particularly if multihoming is costly (in price and/or other ways) and there is limited, or no, interoperability, strong network effects will tend to push the market to one, or perhaps just a few, metaverses. The reason is simple: if you are choosing between two metaverses, say A and B, even if you like A better because of, say, its graphics and interface, if B has far more users than A, you are likely to find B more appealing overall, since you'll have a substantially greater ability to interact with others.

This isn't to say that the existence of network effects necessitates there being just a single Metaverse. It could be that network effects diminish with volume. For example, you may find a metaverse with 10 million users to be similarly attractive to one with 100 million users, all else being equal. In such a case, both the smaller and larger metaverses could simultaneously exist. This scenario is familiar, as we see something similar when it comes to social media (Facebook, X, Instagram, TikTok, etc.) and even operating systems (Windows, OS X, Linux, etc.).

The final demand-related factor we highlight as influencing the number of metaverses is taste heterogeneity. If consumer tastes for Metaverse technology—in terms of, for example, design and interactive features—vary greatly, this could lead to multiple metaverses, where each is built to accommodate different tastes. The likelihood of such an outcome is greater if interoperability is difficult, making it less likely that a single Metaverse emerges that is comprised of distinct, interoperable virtual worlds that each cater to different consumer tastes. On the flip side, if tastes for Metaverse technology are more homogenous (i.e., we all can roughly agree on what we like and don't like about Metaverse design, features, etc.), this would push toward a single Metaverse that accommodates those shared tastes.

Are we headed toward one Metaverse overlord, a few, or control by the masses?

A future of a combined omnipresent Metaverse on the order of the OASIS from *Ready Player One* is unlikely anytime soon, as discussed further in Chapter 9. Meta certainly has the drive to create such an all-encompassing digital environment and has invested more than $36 billion in its Reality Labs unit since 2019.[28] However, the challenges of driving this technological, social, and economic shift are significant, and even if it were successful, it could give rise to the same antitrust concerns that we are seeing in a variety of Big Tech contexts. A key component is cross functionality, including a shared operating system that has proven to be integral in the realization of global internet platforms and shared conceptions of cyberspace.[29] Yet this goal has been elusive to date. One 2022 study, for example, did not find that any of the self-proclaimed Metaverse leaders met all the criteria of a truly persistent, interoperable, multimodal, and shared virtual space.[30] In many ways, the number of metaverses and the number of "overlords," that is, entities that have authority over the operation and activity within them, will have a symbiotic relationship, where each impacts the other. However, as is often the case in economics,[31] we expect much of this two-way relationship to be such that the determinants of the number of metaverses (discussed in the previous question) are key drivers of the number of overlords.

To begin, suppose technological/cost and demand factors are such that the market tends toward a single Metaverse emerging. Such factors include those discussed in the previous section: economies of scale, costs of interoperability and multihoming, multihoming preferences, network effects, and Metaverse tastes. Some of these factors can be described as a barrier to entry, defined as any factor that can prevent newcomers from entering a market or industry, and thus limit competition.[32] High fixed costs, high costs of interoperability and/or multihoming, consumer aversion toward multihoming, and strong network effects all qualify as barriers to entry. Broadly speaking, high barriers to entry can push a market toward having few competitors. Other examples include high start-up costs, established regulation, and consumer brand loyalty.

In the early stages of Metaverse development and competition, the lack of substantial regulations and established competitors (and thus limited brand loyalty and network effects) suggest lower entry barriers, which plausibly has attracted individuals, start-ups, and other creatives to enter in various ways.[33] On the other hand, executives have cited the uncertainty about the return on investment and implementation of technology to be a key barrier to entry in the Metaverse space.[34] Regardless, there have been several early Metaverse platforms with early users spread across them, including Decentraland, Fortnite, Minecraft, Roblox, and The Sandbox.[35]

While the technology itself continues to develop alongside the list of major players in the industry, many businesses are scrambling to be early suppliers in a potentially highly lucrative market.[36] Being a first mover in the industry for the Metaverse can be highly consequential; barriers to entry can materialize as early movers establish brand recognition, a robust, loyal user base, and a large enough customer base to generate significant network effects.

If a single Metaverse does emerge, it could, in principle, be controlled and run by the government in a way not dissimilar to China's ambitions mentioned in Chapter 1. However, at least in the United States, such an outcome is highly unlikely. A more realistic possibility is a single Metaverse with heavy government regulation. If a market has (typically cost and/or demand) features such that it tilts toward monopoly, it can be the case that a single firm operates in that market but with significant government regulation over prices, services, etc. A possible scenario would be that a single firm operates the Metaverse, subject to regulation, but "in exchange for these restrictions," is also granted exclusivity—that is, it would be protected from entry by competing firms.

A final single-Metaverse scenario would be one with a single "overlord"—one firm that operates and controls it—and minimal government regulation. Such a scenario would raise classic monopoly concerns, such as high prices or reduced quality and innovation. Whether such concerns materialize will depend on two types of threats to the incumbent Metaverse firm. One is the threat of regulation: if it becomes clear that prices are exorbitant and/or quality and innovation are suffering due to a lack of competition, it is likely the government will impose regulations (meaning a shift to a heavily regulated monopoly) or attempt to break it up into multiple firms through antitrust law enforcement. The other is the threat of entry by competitors: for example, if fixed costs are high enough to create significant economies of scale, but not extremely high, other firms may find it attractive enough to enter if the incumbent monopolist is making very high profits (e.g., via high prices). If this threat exists, the incumbent monopolist will have an incentive to keep prices low enough (and/or keep quality high enough) to prevent such entry, known as "limit pricing."

The emergence of several metaverses is also possible, which could take several forms of governance. For starters, several metaverses with one "overlord" firm is unlikely, even if consumers have highly heterogeneous tastes (creating an incentive to build multiple virtual worlds to accommodate those tastes). This is because a single firm would have great incentive to minimize interoperability costs, preserving the ability of its customers to navigate across virtual worlds and, thus, resulting in a single Metaverse with multiple, interoperable virtual worlds.

A scenario with several metaverses (at least two), each with a separate firm controlling it, would be consistent with early Metaverse development and preceding technologies (e.g., smartphone operating systems). There are several reasons why technology, demand, and regulation may be such that the Metaverse market would have multiple players. These include reduced entry barriers through technological advancements, stringent government antitrust regulation, and evolving consumer demand. We expand on each:

- *Lower entry barriers through technological advancements.* With continuous technological advancements, VR/AR technology is becoming more accessible to a broader consumer base. This expansion in accessibility holds the promise of a larger market, which would encourage a growing number of suppliers to enter the industry. In fact, in recent years, the industry is already believed to have an entry barrier lower than ever.[37] Similar to the hardware industry, the software industry, like digital technologies, likely will also experience lower fixed costs of production and hence a lower entry barrier over time.[38] Reasons behind it include more accessible knowledge for the industry and lower production requirements.
- *Stringent government antitrust regulation.* As mentioned in the previous section, several large tech firms (Meta, Apple, Google, etc.) have begun the process of entering the Metaverse market. These firms already have attracted antitrust

scrutiny in other markets, and any sign of early dominance of the Metaverse will likely do the same. Any antitrust interventions will counteract market concentration (i.e., control by a few firms or just one firm).[39]

- *Evolving consumer demand.* In recent years, the population has become increasingly diverse, with millennials and Gen Z accounting for a significant portion of the demographic landscape, two of the most diverse generational cohorts in US history. This diversifying consumer base will create diverse needs, presenting opportunities for suppliers to offer tailored products and services in the Metaverse market.[40] As mentioned earlier, greater heterogeneity in tastes would push the market toward multiple metaverses, particularly if interoperability is costly.

We conclude this answer by considering the possibility of a Metaverse, or metaverses, run by "the masses," that is, in a highly/fully decentralized manner. Such a possibility fits within the broader set of pushes toward decentralization, such as what we observe in cryptocurrency (decentralized currency) and even the conceptualized Web 3.0 (a decentralized Internet), both relying on what is known as blockchain technology. It remains to be seen how successful either of these existing endeavors will be, but particularly if Web 3.0 comes to be, the possibility of a decentralized Metaverse increases dramatically. The pace at which these other decentralized technologies are proceeding makes a decentralized Metaverse seem unlikely, but certainly possible in the longer term.

How will the Metaverse and physical world coexist?

There are three basic ways that the Metaverse and physical worlds will coexist with respect to the products and services people buy and consume in each—as complements, substitutes, or (largely) independent. Let us start with complements.

By definition, two products (or services) are complements if a decrease in the price of one increases the demand for the other.[41] Put another way, two products are complements if consumption of one increases the appeal/value of the other. A classic example is beer and pretzels; for many people, a beer is more enjoyable when consumed with pretzels and vice versa.

During the early stages of the Metaverse, firms began identifying and building out products and services in the Metaverse that are complements to existing physical world counterparts. For example, McDonald's has indicated that they are in the process of developing "virtual restaurants," and plan to sell virtual goods (such as branded clothing items for Metaverse avatars) within the Metaverse.[42] These virtual restaurants would offer users the opportunity to place a McDonald's order in the Metaverse which would then be prepared by a brick-and-mortar restaurant and delivered to their location in the physical world.[43] Hence, McDonald's early Metaverse business strategy involves complementary products and services, as they are dependent on the real-world McDonald's business operations.

As another example, the popular gaming platform, Roblox, has generated sponsorships and advertising opportunities with brands like Nike and Gucci.[44] Nike sponsored the creation of "Nikeland" within the Roblox platform, which leaders at the company have said could be a way to test new products and enhance brand recognition.[45]

Through these and other complements in the Metaverse, consumers also get greater value from their physical world counterparts. Perhaps one of the clearest examples of the benefits of complementary Metaverse products and services has already started to manifest in education products.[46] For example, students may find an in-person course on Greek history valuable, and, separately, they may find an immersive Metaverse experience in ancient Greece valuable. However, they may find the combination to be even more valuable than the sum of the two parts; the course is more valuable with the immersive experience, and the immersive experience is more valuable in conjunction with the course.

Two products (or services) are substitutes if an increase in the price of one increases the demand for the other.[47] Conceptually, two products are substitutes if consumers tend to buy one *or* the other for a given purpose. For example, we might think of coffee and tea as substitutes, where consumers are seeking a warm drink and will choose only one or the other; they buy coffee instead of tea and vice versa.

Substitution between the Metaverse and the physical world is a natural extension to online/offline substitution (i.e., substitution between products in the physical world and online) that was first identified in the early days of the Internet.[48] For example, computer manufacturers, such as Dell and Gateway, began offering customizable computers online, which competed with in-store computers produced by HP and others.[49]

As Metaverse technology develops, a new set of consumption choices becomes available to the average consumer. Consumers now have the option between certain physical world experiences, as well as the option to experience some of those activities in a virtual world. For example, an individual might be deciding between going to see a movie or going on a hike. Before immersive Metaverse technology emerged, the only option was to experience these things in the physical world. However, now with the advent of this new technology (particularly in the VR space), this individual could also choose whether (s)he would prefer to watch that movie in VR in the Metaverse, or whether (s)he might prefer to take a hike in the Swiss Alps using a VR headset.

Thinking more broadly, all of us have our own preferences over various products and services (including experiences) that we can consume in the real world or the virtual world; these preferences guide the purchases we make.[50] Much like your choice of what to eat for lunch, fundamental economic principles tell us that you will choose whether to consume in the real world or the Metaverse based upon what option(s) you prefer most, considering everything including prices. Introducing substitutes can make people better off, not just from competition lowering prices but also enhancement in choice.

In the earliest onset of Metaverse technology, we can observe the popularity of certain Metaverse products and see the emergence of real-world substitutes over time. For example, some of the most popular early innovations from Metaverse technology are gaming platforms, and individuals have shown a particular propensity to shift gaming entertainment consumption to the Metaverse.[51] Another example of an emerging substitute is movie watching. The immersive capabilities of VR and AR headsets have the potential to revolutionize the movie industry. In contrast to watching a move in the cinema or on a TV screen, VR/AR technologies enable users to "feel as though they are truly present in a virtual environment, [which] creates a sense of presence and immersion that can stimulate the senses and trick the brain into believing that what they are experiencing is real."[52] There is also potential for consumers to switch to the Metaverse for product demonstrations (demos). Before the advancement of Internet technology, a consumer needed to go to the store to view or try the product before purchasing, no matter if it was a new jacket, new furniture, or an apartment. The Internet has already revolutionized the market, but the Metaverse could take it further. VR/AR technologies will create more realistic 3D models, and thus a more interactive experience.[53] Imagine being able to tour an apartment from thousands of miles away, yet as if you were in the room.

We conclude by noting ways that the Metaverse likely will expand consumption, with new products that aren't clear complements or substitutes for existing physical world products (hence being largely independent). That is not to say that no substitution will occur. If consumers start spending more time and money in and on the Metaverse, it must come from somewhere—time spent doing other things and money spent (or possibly saved) on other products. However, for some products, this shifting of time and money to consume them in the Metaverse doesn't come at the expense of any obvious substitute(s). In addition, to the extent that Metaverse products compete well with physical world products, such competition can drive prices down, leaving consumers with more purchasing power to buy additional products in the Metaverse, but not at the expense of any physical world substitute(s).

Much of the novelty in Metaverse products will come from the immersion—often visual—into digital spaces. For example, running around as a virtual gorilla in the popular game "Gorilla Tag" involves an immersive visual experience both in terms of how you see yourself and how you see your movements translate into what you see. Such a product does not have a clear real-world substitute. Another related example concerns 3D maps. In particular, software developers have gained access to Google Earth's 3D maps, which marks another step closer to an Earth VR app for standalone VR headsets.[54] Such a development will bring consumers more immersive experiences, allowing them to see "a Sim City view of the real world,"[55] which again has no clear, real-world substitute. This will open the market for both navigation services and video gaming.

As one more potentially highly consequential example, consider surgery tools. VR and AR technologies can be used both before and during surgery. In fact, Stanford Medicine already has been using a new software system to create a 3D model of brains that physicians can see and manipulate. As the head of the Stanford Neurosurgical Simulation Lab put it, "It's a window into the brain."[56] Much more than just a new way to stimulate the senses, these technologies have the potential to save lives in ways for which there aren't any clear, real-world substitutes.

3

SPEECH AND EXPRESSION IN THE METAVERSE

Free speech isn't just about speaking. It is also about listening.

~Tim Cook, Apple CEO[1]

The Metaverse could grow into a vibrant platform for creative expression. As some technologists describe, the Metaverse will allow users to craft immersive experiences, collaborate in virtual workspaces, and host virtual events. However, the same freedom that allows for boundless creativity also opens the door to potential misuse. Imagine stumbling into a virtual comedy club where performers blindside an unsuspecting audience—including children—with explicit content.

Picture a political rally gone wrong in virtual reality - what starts as passionate debate descends into a mob of avatars attacking each other. Unlike Twitter fights, these confrontations might feel viscerally real. When does heated political theater cross the line into harassment? Should virtual crowds have the same right to assembly as physical ones?

We can't eliminate all risks from virtual interactions, just as we can't make physical spaces completely safe. But we can think carefully about designing virtual spaces that encourage healthy discourse while protecting users from genuine harm. In this chapter, we'll learn, perhaps unsurprisingly, tha existing laws aren't well suited to address such problems. The law is rooted in physical reality, where the lines between speech and action, between real and pretend, between public spaces and private ones, are clear. The Metaverse blurs all these boundaries.

What opportunities and challenges does the Metaverse present regarding speech and personal expression?

The Metaverse transforms how we express ourselves by offering three key innovations: immersive environments that can make digital interactions feel physically real, customizable avatars through which we can control how others perceive us, and virtual spaces that influence where and how we gather. Each of these elements reshapes the boundaries of human expression. Let's begin by exploring the concept of immersion.

Immersion in virtual reality is the sensation of being physically present in a virtual world. The sensation is more than merely seeing an environment in 3D—it's a more holistic experience that leads one to feel on a more visceral level that they are somewhere else. You are fully aware on an intellectual level that what you're experiencing is an illusion, and yet the technical details seem to fade away when you move within a virtual space where everything you see matches your movements perfectly.[2] Sound deepens the deception. Recent advancements in spatial audio allow users to perceive sound direction and distance with startling realism.[3] Imagine playing a round of virtual mini golf with your friends, and being able to participate in a group discussion where you can turn your head to focus on different speakers—just as you would in the real world.

Perhaps the most intriguing aspect of immersion is the phenomenon of "embodiment" - a sort of deep mind-body connection that some users of virtual reality have reported experiencing.[4] Embodiment occurs when your virtual body feels like "you" - virtual limbs move with the same instinctive grace as physical ones, and actions flow as naturally as they do in the physical world. This can make VR experiences more engaging for users and, as a result, more successful for Metaverse companies.

The power of immersion is a double-edged sword, however. The same factors that make it more compelling than traditional social media or gaming experiences can amplify negative encounters.[5] When harassment occurs in VR environments – as recently reported in several troubling incidents within Meta's Horizon Venues 00 the impact might be even more profound than similar behavior on social media or gaming platforms.[6] The visceral nature of VR means that verbal abuse and hostile behavior can inflict genuine psuchological harm, particularly to female users, who have sometimes resorted to concealing their identities to avoid harassment.[7] This heightened emotional impact of harassment has unsurprisngly deterred user engagement in early Metaverse spaces. For the Metaverse to thrive, it must feel safe.[8],[9]

Meta's reponse reveals what a tough problem this is. Their new "personal boundaries" feature allows users to create an invisible force-field that renders other avatars mute and at arm's length. While this measure offers some welcome protection, it places the burden of safety onto potential victims. Rather than addressing why some users choose to harass others in virtual spaces, this approach asks the targets of harassment to maintain vigilance.

The challenges of immersion extend beyond managing direct harassment. The immersive and embodied nature of the Metaverse also introduces a risk of psychological manipulation.[10] Recent studies suggest that the immersive nature of the Metaverse, combined with the possible ability to track users' emotions and behaviors, could lead Metaverse users or other designers of Metaverse environments to influence the way users think.[11] This could include, for instance, making users more susceptible to persuasive ads, political propaganda, or disinformation Imagine an AI system that learns which emotional buttons to push, adjusting how it persuades users moment by moment.[12] To prevent these forms of harm, content moderation is essential and will be discussed later in the chapter.

Yet, for all its potential risks, immersion also offers unique and powerful benefits that go beyond simply making the Metaverse more fun. Research has shown that virtual reality experiences can significantly enhance empathy toward vulnerable groups—something that the non-immersive world of social media has been noticeably lacking.[13] By allowing users to virtually "walk in someone else's shoes," VR experiences about refugees, for example, have demonstrated a remarkable ability to increase understanding and compassion. This empathetic connection, facilitated by the immersive nature of VR, holds immense potential for bridging divides and fostering a more understanding society.

VR is also transforming the treatment of Post-Traumatic Stress Disorder (PTSD). By creating controlled, immersive environments customized to individual traumatic experiences, therapists can help patients confront and process their trauma in ways that were previously impossible. The visceral nature of VR – presented earlier as a source of liability – can actually optimize some kinds of mental health treatments.[14] The technology has also proved effective in dreating dementia. Elderly patients can step into versions of their past lives – perhaps the street they grew up on, or the dance hall where they met their spouse. These kinds of journeys, which are called "reminiscence therapy," have been used as a kind of virtual time travel with beneficial outcomes.[15]

Immersion could also make education in the Metaverse more compelling and effective than in offline environments.[16] In a series of aspirational ads, Meta has suggested history classes where students can walk through accurate recreations of ancient cities and science lessons where complex molecular structures can be manipulated by hand.[17] The potential for more engaging, effective, and memorable learning experiences seems immense.

As online spaces become more immersive, avatars—virtual representations of users, explored in Chapter 5—are growing as mediums for identity exploration and self-expression. Unlike static profile images, avatars are dynamic, detailed, expressive stand-ins that influence both how users are perceived and how they perceive themselves. As with immersion, this aspect of the Metaverse carries risks and potential benefits.[18]

When people create avatars, they often reveal parts of themselves they keep hidden in physical life. Studies indicate that the process designing an avatar can be a creative way for some people to share their experiences - like making a kind of mask that reflects parts of who they are.[19] More fascinating still is how these digital selves loop back to influence their creators. People who inhabit confident avatars begin standing taller in real life.As one author has explained, "[as] an extended self, the avatar-persona becomes interwoven with how we think of ourselves—it is both object (mine) and me."[20] Avatars, thus, may become more than just digital representations; they could be extensions of the self that mold how individuals see themselves and act. Nonhuman avatars could enhance speech and expression even more. Imagine being able to represent yourself as a geometric form, or a being made of pure light. In the Metaverse, one could even change how their voice sounds to others. Physical markers like race, age, and gender might fade into the background. Some experts believe that nonhuman avatars can enable users to focus on intellectual contributions rather than physical appearances. In a way, this could make the Metaverse more inclusivity.[21]

MIT researcher D. Fox Harrell uncovers the quiet prejudices hidden in avatar design. For instance, the developers of a video game or online space may disasvantage avatars or characters of certain races or genders. Harrell's solution to these problems is an experimental platform he calls "Chimera." Characters within this system navigate social situations that shift based on who – or what – they appear to be, Some players discover what it means to lose privileges they took for granted. Others find their virtual experiences eerily echoing real-world challenges they face daily. Through these digital experiments, Harrell shows how thoughtful design could help virtual worlds become testing grounds for social change rather than just recreating society's blind spots.

Avatars offer a chance for free expression but also harbor risks, especially with the anonymity they provide in the Metaverse. This anonymity, while fostering open dialogue, can unfortunately pave the way for harmful behavior. Unlike social media that encourages authenticity, the Metaverse allows users to adopt constantly changing identities. This flexibility, combined with the Metaverse's live and interactive environment, might lead to lowered inhibitions and the abusive behaviors highlighted previously. The potential for users to create false identities by altering their appearance and voice is a potential concern. This could lead to a surge in verbal abuse or simulated physical attacks. It's important to be aware of these risks and take necessary precautions to prevent such behavior.

The anonymity of Metaverse avatars poses risks that traditional social media never faced. While Facebook and LinkedIn push users toward real names and authentic profiles, the concept of the Metaverse thrives on fluidity – it's a realm where someone might appear as a dragon-slayer one moment and a corporate executive the next. This freedom to shape-shift opens creative possibilities, but also could embolden bad actors. A user's voice modulator and carefully crafted avatar might express their truest self, or serve as armor for launching verbal assaults. Some users might slip into their digital skins and find their usual social constraints melting away, especially in the heat of live interactions. The distance between typing an angry comment and screaming at someone could become surprisingly thin. These aren't just hypothetical dangers - early Metaverse platforms already struggle with users who exploit anonymous identities to harass and intimidate others.

Fortunately, it might be possible to balance avatars' benefits with their risks. Implementing a pseudonymous system where users have consistent but not fully identifiable personas—like those in online games—shows great potential. This setup encourages users to develop characters they grow attached to, investing time and, possibly, money in their creation. Such investment promotes responsibility, as actions have real consequences like losing access to the platform. Studies indicate that pseudonymous users tend to act more honestly and cautiously, reducing harmful behaviors.[22] By adopting pseudonymity, Metaverse designers can foster a safer, more accountable online community, ensuring free expression while curtailing abuses.

Now let's turn to the impact that virtual envronments might have on speech and expression. Metaverse environments, which can range from from fantastical landscapes to realistic simulations, have the potential to open unprecedented avenues for expression and speech. Yet, this limitless potential also comes with its own set of challenges. Privacy, content moderation, and the psychological impact of these immersive spaces are crucial considerations.

Digital architecture has the potential to shape human behavior in ways no website or app ever could.[23] Metaverse spaces can respond to our movements with subtle precision - footsteps can echo, shadows can shift, and objects can have weight and presence. Our brains, finding these cues familiar, may comfortably settle into these artificial worlds. The accumulation of small and subtle environmental details can have a profound impact when added together. When digital objects start casting shadows across your kitchen table, something shifts in how you think about online space. You stop being a visitor and start being an inhabitant.

Virtual environments are proving uniquely effective for people who face challenges with traditional face-to-face communication. For example, in a recent study involving children with Autism Spectrum Disorder (ASD), a collaborative virtual environment (CVE) was used to promote natural communication and cooperation between users through interactive games.[24] Similarly, virtual environments have been used for education and support sessions for individuals with type 2 diabetes, where different types of support were exchanged, including informational and emotional support.[25] Other studies have shown the benefits of thoughtfully designed virtual environments in facilitating interactions among people with chronic illnesses and children on the autistic spectrum.[26] Metaverse platforms seem particularly good at fostering intimate, reciprocal communication.[27]

Metaverse environments might become canvases for expression, as illustrated by artist Sebastian Errazuriz's protest against augmented reality (AR) commercialization.[28] After Snapchat announced a partnership with sculptor Jeff Koons to use AR for placing his artworks in famous landmarks, Errazuriz "graffiti-bombed" the project in protest. By overlaying graffiti on an AR version of Koons's "Balloon Dog" in Central Park, Errazuriz protested what he called "an augmented reality corporate invasion." This act sparked a widespread public debate on virtual rights and property. It also underscores the evolving challenges of virtual ownership and expression, highlighting the need for legal frameworks to adapt to digital realities. Errazuriz's protest serves as a pioneering example of virtual expression and protest that Metaverse environments could enable.

The Metaverse holds boundless potential for creativity, communication, and self-exploration. Yet, navigating its opportunities requires addressing inherent risks with foresight and care. Metaverse platform operators will need to utilize strategies that consider user safety, perhaps by establishing personal boundaries and encouraging pseudonymous interactions, as well as the fostering of a culture of accountability. Moderation practices may also need to be developed. By embracing these challenges with care, Metaverse platform designers can unlock the Metaverse's promise for groundbreaking, inclusive communication and collaboration. The true success of the Metaverse will hinge on immersion that supports positive human connections.

Could the Metaverse face the same content moderation struggles that have plagued social media platforms?

The Metaverse, with its immersive virtual worlds and real-time interactions, presents unprecedented challenges for content moderation.[29] Unlike traditional social media, where monitoring text, photo, and video uploads suffices, the Metaverse demands oversight of complex, live, multimodal interactions—from lifelike avatar conversations to simulated physical touch. This amplifies existing moderation issues and introduces new ones: How do you moderate a virtual protest that turns hostile? How about a user who conveys hate symbols through their avatar? These challenges are compounded by the Metaverse's global nature, where cultural norms can vary. Moderation has never been as complex or as crucial.[30]

Content moderation will likely become a difficult balancing act. If moderation is too lenient, Metaverse platforms risk making harassment commonplace, pushing users away; if moderation is too strict, the platforms will likely be criticized for violating users' privacy and ability to express themselves. Finding the right equilibrium between user safety and freedom of expression is crucial for the long-term viability and appeal of Metaverse platforms.

The global nature of the Metaverse introduces significant challenges in developing universally accepted moderation policies. Because users from diverse cultures will be interacting in shared virtual spaces, creating a one-size-fits-all set of rules becomes increasingly complex. Moderators will need to navigate the nuances of cross-cultural communication, often facilitated by real-time translation technologies, adding another layer of complexity to their task.[31]

For example, a gesture considered friendly in one culture might be offensive in another. Humor and sarcasm, already difficult to moderate in text-based communication, become even more challenging when expressed through avatars in a 3D space. Moderators will need to be culturally knowledgeable. This could mean that Metaverse providers will need to employ a culturally diverse, global team of moderators to effectively manage these cross-cultural interactions.[32]

As discussed earlier in this chapter, Metaverse technologies, such as lifelike avatars, haptic feedback, and spatial audio, create deeply immersive experiences.[33] However, these technologies also increase both the volume and types of content requiring moderation. The ability to simulate physical touch, for instance, introduces entirely new categories of potential misuse that moderators must be prepared to address.

Consider the implications of haptic technology that allows users to "feel" virtual objects or even other avatars. How does one moderate unwanted or inappropriate "touch" in a virtual space? What about the use of spatial audio to whisper harassing comments that only the target can hear? These scenarios present unique challenges that go beyond traditional content moderation.

Moderation could be further complicated by the fact that not all interactions in the Metaverse will be necessarily saved or recorded. If so, this transient aspect of virtual interactions would be a key difference from traditional social media. In traditional online settings, moderators can retrospectively review and address content; but in the Metaverse, many exchanges might effectively vanish before they can be evaluated.[34] This suggests that real-time moderation won't just be preferable, but will in fact be the only viable option for effective oversight.

This transience raises important questions: How can platforms ensure user safety in real time without infringing on privacy? What technologies or strategies can be employed to detect and address harmful behavior as it occurs, rather than after the fact? (Spatial audio in virtual environments could restrict moderators from hearing all nearby conversations, and the vastness of virtual worlds limits their visual oversight[35]). Some experts have posited that these challenges may border on the impossible, given the volume and complexity of these interactions.

AI-driven content moderation is a possible solution.[36] However, while adept at real-time interception of harmful content, AI often struggles with contextual understanding.[37] Consequently, humans might need to remain in the moderation loop.

This points to yet another question: who should have the power to moderate content in these virtual worlds? For the past two decades, a handful of companies in Silicon Valley have played this role across social media, drawing all the blueprints for what can be said and shared online. This centralized approach gave us consistency, but at a price - local communities often found their needs overlooked by distant decision-makers. There's another way to do this, though. Moderation could be managed in a decentralized fashion by users themselves.[38] This approach - embodied today by "the Fediverse" - raises complex challenges of its own, though: When every community can set its own rules, how do we prevent our digital world from fracturing into isolated enclaves? How do we balance local autonomy with the need for some universal standards? The answer might not lie in choosing between centralization and decentralization, but in understanding how to blend them effectively - much as physical cities function through a complex interplay of local councils and central authorities, perhaps the Metaverse needs a similar balance to preserve both community autonomy and collective safety.

Yet another challenge will be balancing universal content policies with diverse cultural norms and legal frameworks globally. This will be critical to maintaining inclusive spaces that enable cultural exchange. A risk in this area will be avoiding a "lowest common denominator effect" in moderation policies, which could either limit expression or lead to fragmented standards.

When people meet in the Metaverse, they leave geography behind but bring their legal systems with them. Picture a virtual gallery opening where visitors arrive from Tokyo, São Paulo, and Berlin - each under different laws about privacy, speech, and digital rights. These overlapping jurisdictions raise questions about how global virtual spaces can respect and follow local laws. Answers might come from unexpected places: cross-border telemedicine platforms have already pioneered ways to navigate different national regulations while maintaining consistent care standards. Their success suggests that the Metaverse could embrace both local governance and shared international principles, rather than forcing a choice between them.[39]

Ultimately, handling content moderation in the Metaverse will probably require collaboration between the people and institutions in power. Platform creators will need to refine their moderation methods to balance freedom of expression with harm reduction. The newness of these challenges also necessitates government involvement to provide legal guidance and guide moderation standards. Ideally, advocacy groups, tech experts, legal scholars, and users themselves will also have the ability to shape Metaverse policies by offering insights. When these kinds of groups work together - as we've seen in successful online gaming communities - they create spaces that protect free expression while keeping users safe from harassment and harm. The Metaverse's success may depend less on breakthrough technology than on this kind of thoughtful collaboration.

Is the legal system ready for speech and expression in the Metaverse?

The Metaverse integrates digital, augmented, and physical worlds to create immersive, interconnected virtual environments for activities from fantasy gaming to remote work. As these spaces evolve, lawmakers will likely struggle to address speech and expression, especially regarding copyright, trademarks, privacy rules, and access for people with disabilities. For example, will virtual world developers have the right to censor residents' speech or avatar expression? While the goals of current laws will remain relevant, their specifics will need rethinking and adapting to prevent harm and protect rights in these new realities. Let's explore potential legal impacts and modernizations that will be needed.

Copyright and trademark

In the Metaverse, the fusion of creativity, expression, and technology blurs the lines of conventional copyright law. Imagine a scenario where a user recreates a scene from a copyrighted movie in a virtual world. This act prompts vital questions about permitted forms of expression and how copyright laws apply in digital spaces.

"Fixedness" is a key principle of copyright law, requiring that a protected work be captured in a stable, tangible form allowing it to be perceived, reproduced, or shared.[40] While traditional mediums for fixed works include paper, canvas, film, and audio recordings, digital creations in virtual realms often remain fluid and changing. For instance, an interactive virtual sculpture could alter itself indefinitely, never existing in one fixed state. Can such mutable, interactive digital works still meet fixedness criteria for copyright protection? This sort of question may necessitate rethinking what forms of digital expression should qualify as "fixed" works.

Trademark concerns also enter uncharted territory in the Metaverse.[41] Trademarks traditionally protect brand identity in the context of "use in commerce."[42] This term implies that a mark must be used in commercial sales or advertising goods or services to qualify for trademark protection. The focus is on using the mark in the marketplace to identify the source of a product or service, thereby preventing consumer confusion. The Metaverse, however, blurs the lines between commercial and noncommercial use. Is a virtual avatar's use of a trademarked logo, without any transaction or traditional commercial activity, considered "use in commerce"? This ambiguity challenges the application of trademark law, necessitating a nuanced understanding of commerce in digital and non-physical spaces.

Consequently, celebrities like Rihanna, Snoop Dogg, and Miley Cyrus are now filing trademarks to protect their brands in the emerging digital world of the Metaverse.[43] This trend signals that they view these interactive, immersive virtual spaces as a major future market to sell products, market services, and engage with fans. By establishing trademark rights early on, these stars can license their brands for virtual goods, claim royalties from Metaverse concerts and events, or create exclusive social spaces—essentially staking their claim in a new digital frontier.

First Amendment law

The First Amendment shapes how Americans understand their right to speak freely, but it was written for a world of town squares and street corners - not virtual realms.[45] While it stops the government from silencing citizens or favoring certain religions,[44] it says little about what user speech Facebook, Twitter, or Metaverse platforms must publish. This creates a paradox: when you step into a virtual world, it feels like entering a public space. You might meet friends, debate politics, or join a protest rally - just as you would in a physical town square. But these digital spaces, despite their public atmosphere, are more like privately owned shopping malls than public parks. As a result, they are not subject to the same rules regarding freedom of speech. It is understandable that some social media users have been confused by this. After all, what's the difference between gathering in a town square and gathering in virtual reality? Legally, the difference is everything - even if these private platforms have become our new public squares.[46]

As these platforms become more powerful, could they become mini governments bound by laws like the First Amendment's protection of free speech? We are still determining how this might happen or how it would change what content is allowed on the platforms. Additionally, how do we balance the rights of Metaverse companies to control their spaces with the rights of users to speak freely? This might mean finding new ways for users to have a say in the rules and how content is handled. Finally, as the Metaverse blurs the line between virtual and real life, we might have to change how we think of public and private spaces. This has significant consequences for managing what people can say and do in these virtual worlds.

Criminal law and tort law

Criminal law aims to deter harmful behaviors threatening society through penalties targeting offenders. However, this body of law was developed to address problems in the real world rather than in virtual spaces. As a result, the legal framework might need to adequately recognize new forms of harm possible in the Metaverse. For instance, personalized avatars could enable anonymous users to harass others by using extremely abusive speech. Targets might experience nightmares, panic attacks, or withdraw socially after the incidents. However, some legal experts debate whether emotional distress merits criminalization when resulting from "just words" in a digital realm.[47] Determining guilt could also be complicated if offenders come from jurisdictions where verbal attacks do not constitute crimes. How should conflicting cultural perspectives be addressed regarding conduct expectations and punishments in shared virtual environments that are accessible globally?

Cyberbullying and stalking raise similar challenges. Does criminal law apply if users repeatedly threaten and stalk avatar representations of real individuals?[48] Supporters emphasize that harassment targeting embodied digital personas carries similar emotional impact and post-traumatic symptoms as equivalent physical offenses.[49] Yet opponents counter that anonymity makes threats in virtual settings appear less credible. Like criminal law, tort law frameworks dealing with private harms and disputes, like negligence or defamation, grapple with the Metaverse's advent. For instance, expert consensus indicates prolonged immersion in virtual realities could produce physical injuries such as eye strain or repetitive motion symptoms. Should liability apply if device makers fail to warn users sufficiently or design with reasonable safety precautions suited to extended use? Additionally, in interactions between avatars within virtual worlds, expectations around reasonable conduct are undefined—what constitutes negligence? If a user left a virtual trap negligently activated, resulting in emotional distress for an unsuspecting victim, what standards determine the appropriate duty of care or injury compensation, given the acts produced no direct bodily harm? Courts have yet to address such questions, necessitating evolution in principles as technology reshapes human experiences of place, embodiment, and harm outside the constraints of physical reality.

Consumer protection law

Imagine browsing a virtual dealership showing an augmented view of a car model. Sleek curves glisten as embedded videos tout blazing acceleration and agile handling. You purchase the digitally layered vehicle confident in the promised performance capacities. But upon receiving the physical version, you find a lackluster sedan straining up hills and leaning sharply around turns. Does manipulated marketing in the Metaverse qualify as false advertising when coins get collected? Consumer protection laws shield shoppers from deceptive practices by businesses seeking profit through fraud. If products exist purely digitally, do they fall outside of definitions and protections for tangible goods that Metaverse users might expect?

Child protection

Laws safeguarding children from inappropriate content and conduct face complex questions as virtual interaction advances. Metaverse platforms allow users to inhabit immersive, persistent digital spaces through avatars. Here, the potential for minors to encounter mature themes raises concerns.

Traditionally, child protection regulations concentrated on television and print media with centralized distribution amenable to oversight.[50] By contrast, the Metaverse's decentralized, user-generated ecosystem poses novel challenges. The anonymity afforded by avatar identities and vast interconnected virtual terrain impedes monitoring what minors access. The global user base strains consistent enforcement across jurisdictions when content deemed inappropriate in some countries remains permissible in others.

Another quandary regards the immersive medium's psychological influence. Research confirms interactive media's amplified emotional impact relative to passive viewing, signaling heightened harm risks for minors from disturbing immersive online content.⁵¹ Yet current legal paradigms overlook this. Determining accountability also grows complicated as users generate a lot of content.

Defamation law

When someone spreads lies about you at work or in your community, defamation law offers clear protections. But what happens when those lies come from a cartoon avatar in a virtual world? The Metaverse muddles our traditional understanding of reputation and harm. An anonymous user hiding behind a digital mask might spread damaging rumors about your virtual gallery opening or business venture, affecting both your virtual and real-world reputation. Yet proving who they are - or measuring the real damage they've caused - becomes far more complex than in the physical world.

In the physical world, defamation cases follow clear geographic rules - if someone libels you in Cincinnati, Ohio law applies. But where exactly does a defamatory statement "happen" in the Metaverse? If a user in Tokyo damages the reputation of a virtual shop owner from São Paulo inside a virtual mall hosted on servers in Ireland, which country's laws apply? The borderless nature of virtual worlds scrambles our traditional legal maps.

The future of reputation in virtual worlds won't be shaped by technology alone, but by how we adapt our legal protections to these new spaces. Just as early internet law evolved to handle email scams and cyberbullying, our defamation laws must now catch up to a world where reputation spans both physical and virtual realms.

The Metaverse isn't just another social network - it's a fundamental shift in how we gather, speak, and interact online. Social media services have long struggled to moderate content, where interactions are largely photos, videos, and text. Now imagine moderating a virtual protest where thousands of avatars gather in real-time, each capable of broadcasting speech, images, and gestures simultaneously. Our current legal system wasn't built for a world where someone's virtual self might be as valuable as their physical assets, or where "location" of speech becomes meaningless. The Metaverse's architects face a critical choice: either adapt old speech rules to new spaces, or build fresh frameworks that appreciate both the power and peril of immersive communication. This choice will determine whether the Metaverse becomes a vibrant public square or a chaotic free-for-all.

4
SECURING THE METAVERSE

Cybersecurity failures have become an all-too-common occurrence in twenty-first century life, impacting firms, governments, and ultimately people worldwide who become victims of identity theft and ransomware. In 2021, for example, multinational firms averaged $13 million annually in cybersecurity costs, which was nearly double the spending from 2020.[1] Smaller firms are also being hit, with the average cost of a data breach nearing $5 million by 2023.[2] Indeed, this is mirrored by cybersecurity spending at ventures with fewer than 10 employees, increasing almost tenfold annually between 2020 and 2021 to over $12,300 dollars.[3] Globally, McKinsey estimates that organizations are spending more than $150 billion annually on cybersecurity, but that by 2035 total losses may exceed $10 trillion.[4]

But as we think about securing the Metaverse and any persistent, immersive, virtual world, we should consider that this is just one aspect of a larger set of problems introduced here but discussed further in Chapters 5 and 6 (identity and privacy, respectively). This is especially the case given that hacking may be thought of, more broadly, according to the cybersecurity guru Bruce Schneier, as "an activity allowed by the system that subverts the goal or intent of the system."[5] In other words, there are more ways to hack a system than by using an exploit such as a cyber attack to take advantage of a vulnerability and cause a data breach. As we will see in this chapter, in some ways, that is the most straightforward security challenge that the Metaverse faces.

We begin by exploring lessons from notable cybersecurity failures in the history of the Internet before moving on to consider to what extent the Metaverse makes these security challenges worse, and what consumers—and businesses—can do about it.

What lessons should we learn from cybersecurity failures on the Internet as we look to the Metaverse?

Hacking is far older than the Internet. In fact, the term "hack" does not even originate in the computer context, but rather may be traced to MIT's fabled Tech Model Railroad Club circa 1961 when members hacked their high-tech train sets to modify their functions.[6] As Walter Isaacson noted in *The Innovators*, the members "embraced the term *hacker* with pride. It connoted both technical virtuosity and playfulness."[7] Under this framing, there have been a variety of notable hackers throughout history, such as the Comptroller General of the Vichy French Army, René Carmille, who in 1940, was an early "punch card computer expert" and took it upon himself to reprogram the punch card machines to slow down the processing. He thus delayed the Nazis' plans and saved countless lives in the process.[8] Others are the early "phreaks," such as by Joe Engressia who, starting in 1957, used whistles at a particular pitch to fool AT&T's systems to get free, long-distance calling.[9]

Although the first Internet worm, called Creeper, had been developed in the 1970s,[10] the modern connotation of "hacking" in the cybersecurity context became firmly established by the 1980s, with blockbusters like *Wargames* helping to instill the idea in the popular imagination.[11] Despite this, even by the late 1980s, confusion abounded. Consider the acts of Robert Tappan Morris (son of the famous cryptographer Robert Morris, Sr.) who, as a 20-something graduate student at Cornell in 1988, wanted to know how big the Internet was—that is, how many devices were connected to it.[12] He launched a program to aid his study and, in the process, unintentionally created one of the most damaging worms to that point.[13] By the time it was stopped, some three days after its launch, tens of thousands of systems had been infected, and hundreds of thousands of dollars in aggregated costs had been amassed, leading to it being called the "first significant cyber attack" in history.[14] Still, reporters were rather new at covering such an episode, and a few asked whether people could catch the computer infection as though it were the flu.[15]

At its most basic level, the Internet is composed of a series of cables, computers, and routers.[16] Innocent or malicious hardware flaws, along with design decisions, in this physical infrastructure can give rise to myriad vulnerabilities. These take many forms, including spyware, viruses, worms, Trojan horses, and distributed denial of service (DDoS) attacks.[17] There is no panacea for dealing with these challenges. At best, these cyber risks can be managed by deploying a range of frameworks and standards, from specific controls such as multi-factor authentication (the annoying text messages and emails that are used to confirm your identity) to concepts such as zero trust security.[18]

Trust in the context of computer networks refers to systems that allow people or other computers access with little or no verification of who they are and whether they are authorized to have access. Zero trust, then, is a security model that takes for granted that threats are omnipresent inside and outside networks; in other words, why bother locking the door if the burglar is already inside the house?[19] Zero trust instead relies on continuous verification via information from multiple sources. In doing so, this approach assumes the inevitability of a data breach. Instead of focusing exclusively on preventing breaches, zero trust security ensures instead that damage is limited and that the system is resilient and can quickly recover. Using the public health analogy, a zero trust approach to cybersecurity assumes that an infection is only a cough—or, in this case, a click—away and focuses on building an immune system capable of dealing with whatever novel virus may come along. If zero trust security were widespread, the ransomware epidemic would be far less threatening.

But there are at least four main challenges to achieving zero trust security. First, legacy systems and infrastructure are often impossible to upgrade to zero trust. Achieving zero trust security requires a layered defense, which involves building multiple layers of security, not unlike a stack of Swiss cheese. But this is challenging in systems that were not built with this goal in mind because it requires independent verification at every layer.[20] Second, even if it is possible to upgrade, it's going to be expensive. Simply put, it's not easy, it's time-consuming, and it's potentially very disruptive to redesign and redeploy systems—especially if they are custom-made. In persistent, immersive realities, such as the Metaverse, such system downtime would be especially noticeable.

Third, peer-to-peer technologies, like computers running Windows 10 on a local network, run counter to zero trust because they rely mostly on passwords—not real-time, multi-factor authentication. Passwords can be cracked by computers rapidly checking many possible passwords—so-called, brute-force attacks. This makes it all the more important in the Metaverse context to require strong passwords and enable multi-factor by default.[21] Fourth, migrating an organization's information systems from in-house computers to cloud services can boost zero trust, but only if it's done right. This calls for creating new applications in the cloud rather than simply moving existing applications, including those related to the Metaverse, into the cloud.

These issues underscore the fact that creating a zero trust environment, either in the Metaverse or a more traditional network, is no easy feat. Facebook's challenges in this area underscore this point. It is well known that the firm was hit with one of the biggest fines in the history of the US Federal Trade Commission (FTC). The FTC cited lax cybersecurity efforts and involvement in the Cambridge Analytica scandal in which the private information of more than 50 million Facebook users was harvested to influence the behavior of US voters.[22] Since that decision, Facebook (now Meta) may have taken corrective measures; nonetheless, the scandal broadly highlights challenges in determining which company(ies) to trust with fashioning an immersive reality experience for millions and which will treat user cybersecurity and privacy as its corporate social responsibility.[23]

Focusing on the Metaverse in particular, a range of cybersecurity vulnerabilities are glaring. These include challenges of phishing, supply chain attacks, and fake identities (a topic that is discussed further below).[24] In particular, cybersecurity measures are needed in the Metaverse to help safeguard non-fungible tokens (NFTs), which are built on blockchain technology. The tamper-proof power of blockchains—so long as no single entity controls more than 50% of the computing power on the network— is powerful given the extent to which cryptographic principles, which are designed for information security, are also "paradoxically . . . a tool for open dealing."[25] Further, the smart contracts that enable NFTs can be vulnerable to hacks, leading to some $70 million in reported losses.[26] Plus, there's the larger issue that the platforms are the ultimate owners of NFTs and other virtual assets on their platforms through an examination of their terms of service.[27]

As discussed further in Chapter 8, the rise of generative AI will also make cybersecurity in the Metaverse that much more problematic given that it is an environment ripe for abuse, including with regard to disinformation and deep fakes. More robust visualization techniques could help address some of these cybersecurity concerns,[28] as would hardware engineered with security in mind from the outset.[29] Similarly, encryption and multi-factor authentication is just as important in the Metaverse as on the Internet.[30] Dedicated security frameworks and initiatives tailored to the unique threats posed by the Metaverse would also be instrumental in making meaningful progress toward addressing these threats, which could be modeled on Center for Internet Security (CIS) controls, the National Institute for Standards and Technology (NIST) cybersecurity and privacy efforts, or the International Standards Organization (ISO). Industry could also collaborate, as is already happening in the AI context, to instill cybersecurity norms in the same vein as the Siemens Charter of Trust and the Cybersecurity Tech Accord. But these will not solve the cybersecurity woes facing the Metaverse, which will only continue to evolve as we live more of our personal and professional lives online.

If we live more of our lives online in persistent virtual communities, aren't we just making ourselves more likely to be targeted by cyber criminals?

If you've never done it before, it's worth trying an experiment—download the data that a major platform like Google, or Facebook, has amassed on you over the years. It's (relatively) easy to do so.[31] One of the authors did so in 2023 and found some surprising results. For example, unless you opt out, Google tracks the apps you use, your location history, and your search and YouTube viewing history.[32] Facebook tracks your likes, networks, location, and potentially much more, and here's the kicker—even if you don't have a Facebook account, you may still have a "shadow" profile if your friends use the service.[33] Such tactics are par for the course in the United States, which does not have omnibus privacy protections, like Europe and much of the rest of the developed world, though an increasing number of states are starting to change that with more comprehensive state-level laws.[34]

Privacy is addressed further in Chapter 6 but, suffice it to say, such patchwork protections likely are driven, at least in part, by user acquiescence on the topic. For example, according to Pew data from 2018, even as 88% of US respondents aged 18–29 and 78% of those aged 30–49 use social media, only 9% of these users said that they were "very confident" that the firms would adequately safeguard their data.[35] What is more, 62% of Americans believed that it was not possible to go about their daily lives without having their data collected, while more than 80% felt that they had little to no control over how their data was used and shared.[36]

Such user discontent is, in many ways, unsurprising. By one estimate from the Privacy Rights Clearinghouse in 2022, there were more than 540 data brokerages operating in the United States alone that are in the business of repackaging private data on millions of Americans and reselling it to the highest bidder.[37] The US FTC estimates that each such broker has on average more than 3,000 "data segments" for the average US consumer that can range from shopping habits and purchase history to phone numbers and home and email addresses.[38] We contribute to this digital footprint oftentimes without even realizing it. But every click, Yelp or Airbnb review, posted comment on your local newspaper's site, and like on a social media site makes your footprint that much bigger and more valuable to advertisers.[39]

Individually, each data profile may not be that lucrative. The *Financial Times* has, since 2013, offered a calculator permitting anyone "to check the worth" of their private information to shed some light into the opaque data brokerage industry.[40] Although its methodology may not be ironclad, it does shine a light on how owning a home, getting married, and having kids raises your value given your associated spending habits. Nevertheless, individually it's quite low; in the case of one author, just forty cents. Yet although the overall value of this industry is difficult to calculate (but is likely massive) it has been reported that an average email address is worth $89 to a brand over time. In 2012, for example, the data brokerage industry posted some $150 billion in revenue[41]—a market total that more than tripled to some $450 billion by 2022.[42] Collectively, our data is worth trillions, since it is powering the growth of some of the largest firms in the world, including Alphabet and Meta. These firms, simply put, may not exist in their current forms but for America's relatively lax approach to personal privacy protections.

The impacts of the COVID-19 pandemic continue to be felt in so many ways, including with regard to our data footprints, with the average amount of data per individual jumping 150% since the start of lockdowns in 2020.[43] As we spend more of our lives online, it's increasingly the case that "all identity is digital identity," as Eva Velasquez, president and CEO of the nonprofit Identity Theft Resource Center, has said.[44] It is worth noting that, while the Metaverse may make this problem that much worse by providing ever more opportunities for data harvesting in a truly immersive online context, much of the data aggregated on the average user is the result of public records: "voter registration rolls, property and court records."[45] This will not only be a boon to advertisers, but could also raise the risk—and cost—of data breaches and resulting identity theft.[46] By 2021, the FTC already had recorded 1.4 million cases of identity theft, which was double the 2019 number.[47] There's also the concern that the Metaverse could lead to an even greater harvesting of biometric information that can, in turn, also be repackaged and sold to companies like Clearview AI, which has scraped social media sites for images that they have amassed and turned over to law enforcement and government agencies to aid in solving crimes.[48] So far, there is only a scattering of state-level laws like Illinois's Biometric Information Privacy Act, which requires opt-in consent for recording someone's facial image.[49] What is more, the proliferation of AI agents and virtual assets in the Metaverse could make users more vulnerable to scams and disinformation.[50]

It will take a concerted effort, on the part of Metaverse platform owners and regulators, to improve cybersecurity in the Metaverse as users spend more time in such an immersive digital environment. Cyber hygiene training is vital to help protect potential victims of phishing scams and avoid the Metaverse becoming a playground for cybercriminals.[51] Mandating multi-factor authentication and helping users opt out of data collection so as to limit their digital footprints as much as possible is also critical.[52] This will be challenging in the Metaverse context given the extent to which the business model is based on advertising.[53] A scene from the *Ready Player One* movie by Ernest Cline is telling in this regard, in which a cutthroat executive tries to make the Metaverse into a marketing machine by filling the users' visual field with as many ads as possible, up to the point at which seizures are induced. Such worst-case scenarios beg the question of what more Metaverse users can do to help protect themselves and their identities in such an immersive environment.

Can the Metaverse ever be secured?

The rapid development of the Metaverse has presented exciting opportunities for individuals to explore immersive digital realms and for companies to exploit these prospects. In the virtual world, social, economic, and cultural activities are possible in ways comparable to real life, and the number of Metaverse users has dramatically increased.[54] However, it is important to safeguard your online presence, personal information, and protect yourself against online harm as Metaverse crimes such as fraud, theft, harassment, cyber-bullying, and other forms of online abuse increase, something we explore further in Chapter 4. After all, more studies are confirming that harassment in the Metaverse "can have profound psychological impact[s] similar to real-life attacks."[55] This section surveys some essential measures for protecting yourself online and in the Metaverse and offers practical guidance for maintaining privacy, securing data, and staying safe.

How do we implement strong security practices?

Using the Metaverse can expose a user to several attacks, such as fraudulent messaging, social engineering (including copycat/imposter websites and social media accounts), fraudulent emails, fake tech support, and bot-controlled messaging.[56] These fraudulent schemes have been effective in luring victims to click on malicious links or attachments, leading them to divulge personal and sensitive information.[57] In 2022, the server of a metaverse environment with blockchain-supported transactions was found to be compromised. Members were messaged about an exclusive getaway, leading to thousands of digital assets being drained from the wallets of members who had interacted with the hackers.[58] Other common forms of attack include malicious airdrops and giveaways where scammers take advantage of project owners offering their tokens or NFTs to investors. Scammers trick owners into clicking malicious links which would ultimately lead to loss of the victim's digital assets. In one case, hackers used a malicious airdrop phishing scam to steal over $1 million worth of digital assets.[59] In another instance, scammers stole half a million dollars through a phishing campaign that involved obtaining a user's seed phrase through social engineering or copying legitimate websites to gain control of the victim's wallet and digital assets.[60] In another instance, scammers used an ice phishing method by creating fake websites associated with a metaverse environment and paid to have their copycat website appear at the top of search results, thereby tricking unsuspecting people into connecting their wallets and ultimately giving hackers access to their metaverse accounts.[61]

Implementing strong mechanisms, such as two-factor or multi-factor authentication, will help prevent these cyber attacks and scams.[62]

According to the Cybersecurity and Infrastructure Security Agency (CISA), multi-factor authentication (MFA) is a layered approach to securing physical and logical access where a system requires a user to present a combination of two or more different authenticators to verify a user's identity for login. MFA increases security because even if the authenticator is compromised, unauthorized users will not be able to satisfy the second authentication request and will not be able to access the targeted physical space or computer system.[63] Implementing MFA makes it more difficult for threat actors to gain unauthorized access to a company's information systems and facilities, such as remote access technology, email, and payment systems, even if passwords or personal identification numbers (PINs) are compromised by phishing attacks or other means.[64] MFA requires both an account password and a temporary code that is only available on another device, such as your phone. As a result, hackers and cybercriminals will not be able to steal your account or gain access to it unless they have your personal information, such as your password or phone. Picture a stack of Swiss cheese—every security method has holes, but the more layers there are, the harder it is for hackers to find a way all the way through.

Another security measure to observe while online and in the Metaverse is the use of strong and unique passwords. A password manager can help by storing your passwords securely and generating strong passwords. The problem with such services is that all your eggs are in one basket, so when the manager is breached, problems can compound as occurred when LastPass was compromised in 2022.[65]

Staying safe online and in the Metaverse requires updating security tools and programs. To help fix any vulnerabilities that hackers could exploit, individuals should ensure that their data is regularly backed up, the computer's software, antivirus, firewalls, and other security tools are also up to date with the latest security patches.[66] Additionally, using secure and trusted networks and enforcing security by building continuous awareness into these networks are smart practices. Platforms will need to implement visibility and analysis throughout the fabric of the Metaverse to detect anomalies and uncover nefarious activities.[67] Access controls are equally essential, whether risk-based access controls, role-based, discretionary, or other access control types, depending on the organization and its users' needs. These controls should be effective to ensure that users can determine who they are interacting or communicating with and whether they want to allow or deny someone access to applications.[68] Another essential tip to consider is to look at the app permissions and check which permissions are being requested. Any application that requires access to data that is not essential to its operation is a significant red flag.[69]

How can we adopt responsible data and privacy protection practices in the Metaverse?

Cybersecurity is critical to the success of the Metaverse, especially given the high propensity for cyber attacks in these environments. The Metaverse represents a huge potential threat to data and privacy, as every movement of our virtual selves can be recorded, analyzed, and potentially monetized in ways we do not yet fully understand.[70] The use of augmented reality and virtual reality devices increases the impact of a data breach, as these devices collect copious amounts of personal information and user data, thereby increasing the likelihood of being hacked and cyber attacks.[71] When users wear devices like VR headsets, organizations can collect data such as their head, eye movements, or voice, "meaning, within a few seconds, we can identify it is you exactly wearing the device. This is a very serious potential privacy problem for the virtual world" according to Philip Rosedale, founder of Second Life, an online world that allows people to hang out, eat, and shop virtually.[72]

As users explore the Metaverse, the possibility of a major data breach like the 2018 Facebook and Cambridge Analytica scandal looms large.[73] Individuals should understand, review, and comply with privacy policies of websites and online services. They should equally be mindful of personal information that is shared online and in the Metaverse. Online presence should be monitored regularly, and unauthorized use of personal information should be checked frequently.

On the other hand, companies should focus on developing secure software and implementing adequate data protection measures in the Metaverse. Secure coding, rigorous software testing, and appropriate account security should be a priority for companies in the Metaverse.[74] Maintaining digital literacy and understanding the rules of netiquette can help keep one safe while adapting to the Metaverse.[75] Platforms should also create and enforce data accountability and data protection responsibilities in the Metaverse.[76] Rosedale suggests, while using LinkedIn as an example, that users can adopt a "web of trust" to exchange information with others to make it easier to establish trust. He adds that identifying people you trust and sharing information with other trusted people helps gauge whether they have mutual friends in common with new people.[77]

How do encryption and VPNs work?

Another way that users can ensure protection online and in the Metaverse is through encryption, which is a way of scrambling data so only authorized parties can understand it.[78] Data encryption is necessary to prevent data breaches and maintain data integrity.[79] Since vast amounts of personal information are managed online and in the Metaverse, it is necessary to use a system that encrypts data and does not collect information.[80] Platforms may use encryption to secure the transmission of data, such as passwords and transactions.[81] Additionally, users can set up a virtual private network (VPN) to secure their internet connection. A VPN encrypts internet traffic and changes your IP address, preventing tracking of your identity or activity. Third parties and malicious actors cannot track your location in the real world, and VPNs improve privacy and guarantee anonymity when interacting online and in the Metaverse.[82]

How do you maintain adequate protection against cybercrimes in the Metaverse?

Gartner, a research firm, predicts that hundreds of millions of people will spend at least an hour in the Metaverse per day by 2026.[83] This potentially large customer base will equally attract cyber criminals, making it essential to understand the types of crimes that can be perpetrated and how to protect against them. In addition to the crimes like data breach and fraud outlined above, cyber-bullying, harassment, and hate speech are issues against which users may have to protect themselves in the Metaverse. Researchers have shown that VR devices are vulnerable to inference attacks that can reveal private information. A victim's AR device can be tracked in real time, compromising location privacy, and an attacker can cause physical harm through attacks that induce cybersickness and user disorientation.[84]

There are many instances of sexually inappropriate behavior in the Metaverse. In 2021, South Korea's Ministry of Gender Equality reported a sexually explicit act in the Metaverse world.[85] A BBC researcher posing as a 13-year-old girl in a centralized Metaverse app witnessed "grooming, sexual content, racial slurs, and rape threats,"[86] while a 2018 study found that 49% of women who regularly used VR reported at least one instance of sexual harassment.[87] Some tech companies offer tools to prevent and report such abuse. For example, Meta provides a safety measure called Safe Zone. This is a protective bubble that users can activate if they feel threatened. When Safe Zone is in place, no one can touch, talk, or interact with the user until the user signals that they would like the Safe Zone lifted.[88] Individuals (including parents) need to stay updated on these tools to remain safe in the Metaverse and report violations. They should familiarize themselves with the terms of service and community guidelines of the Metaverse and follow the platform rules and policies. On the other hand, platforms should integrate intuitive and accessible corporate safety measures, such as automatic personal distance restrictions (like Safe Zone), universal alert gestures, tutorials explaining the rules, and active moderation.[89] Platforms should also create rating mechanisms for age-appropriate access and use them to mitigate these risks within the Metaverse and online.[90] For example, replicas of the 2019 Christchurch Mosque shooting aimed at very young children have been found multiple times on the Roblox platform, although the company has gone to great lengths to curb such content.[91] A survey conducted by WPP firm Wunderman Thompson found that 72% of Metaverse-savvy parents are concerned about their children's privacy in the Metaverse and 66% are concerned about their children's safety. Therefore, parents and users may wish to take advantage of products created by virtual companies for young people to protect their security and privacy.[92]

Are there any unique cyber threats in the Metaverse?

Nick Biasini, head of Outreach at Cisco Talos, warned that some cyber threats are peculiar to the Metaverse and the technologies that power it, particularly blockchain, cryptocurrencies, and NFTs. He notes that "one of the challenges . . . coming up is going to be related to defending your intellectual property and branding."[93] Intellectual property (IP) is a term that encompasses a wide range of creations, ranging from trademarks and copyrights to patents and trade secrets. In the context of the Metaverse, protectable intellectual property assets range from copyrightable literary works to all trademarks, including brands, slogans, and design patent protection for different configurations.[94] As this new frontier evolves, protecting IP is critical to maximizing the potential return on investment. IP infringement can occur in the Metaverse at every step. For instance, selling copyrighted NFT artwork in the Metaverse may involve counterfeit and authentic artworks. As in the real world, trademark infringement can occur when buying and selling virtual goods. The chances of leaking trade secrets, which are crucial in the Metaverse, are very high given the wide range of potential attacks as Siemens experienced following a data breach.[95] Regarding IP in the Metaverse, it is essential to maintain the same precautions as in real life (IRL). The best way to protect one's innovation in the Metaverse is to register your patents, trademarks, and copyrights with the relevant authorities. This protects your work and rights from infringement.[96] Businesses and individuals should conduct an intellectual property analysis of the Metaverse to identify patented domains and prevent intellectual property infringement. One should also consider creating terms of service agreement that governs using your IP in the Metaverse.[97] Tracking and documenting your portfolio is also important to ensure that all assets are adequately tracked.[98]

Additionally, the Metaverse can pose physical and emotional risks, including anxiety, nausea, and eye strain. To protect and safeguard against these risks, it is vital to take regular breaks, avoid using the headset in dimly lit conditions, ensure there is a clear space, and remain aware of your immediate surroundings to prevent accidents.[99]

In sum, what can I do to protect myself online and in the Metaverse?

Various visions of the Metaverse, such as those portrayed in *Tron*, *The Matrix*, *Free Guy*, and *Avatar*, include attempts by protagonists to protect themselves from the abuses of the system. This, as Schneier has stated, may be thought of as "[a] complex process, constrained by a set of rules or norms, intended to produce one or more desired outcomes."[100] The Metaverse is often portrayed in dystopian terms, as distracting humanity and allowing human bodies to be turned into batteries, a la *The Matrix*, or controlling a local population to enable resource extraction as in *Avatar*. But there are also depictions of users protecting themselves from the system by ultimately taking control of it, as Neo accomplishes at the end of the first *Matrix* movie. A more recent wrinkle is a generative AI character helping to reshape a Metaverse into a more desirable, ethical landscape, as in *Free Guy*. Many of these stories and films, though, contain elements of real danger in the real-world Metaverses that have been created to date, including tracking, that necessitate a look at what regular users who may not be the "chosen one" can do in response.

There are luckily quite a few techniques that may be used to limit your digital footprint and generally be a less tempting target in the Metaverse. The problem is that many of these take some added effort on the part of users; they're not automatic opt out settings, and thus it's easy to forget their importance. These steps include the following:

1. *Deactivate and delete old shopping and social media accounts*: the more sites that have access to your shopping habits and social network information, the bigger your footprint is and the more likely that you'll be targeted by criminals.
2. *Delete old email accounts*: Maybe you have email accounts that you haven't checked in some time and are probably—let's face it—full of spam. One of those accounts may be a Yahoo! email address that you opened more than a decade ago. Since it's been so long, you might think that it's no big deal to let it sit. But the problem is, the data resting on those accounts could be a valuable tool for hackers to learn more about you such as by looking through old email attachments, and gain access to other systems by using the old email address as a means of authentication (especially if you happened to reuse a password). Yahoo!, for example, suffered a massive data breach in which more than three *billion* email addresses were compromised.[101] So, don't just change your password, delete the old account (after migrating any information that you'd like to keep).
3. *Check your privacy settings and disable location tracking*: You should disable location tracking to prevent a service from tracking you across websites or the Metaverse. Sometimes, that's easier said than done. Facebook and Google have faced scrutiny from the FTC over the years for various reasons, including for allegedly violating their own privacy promises.[102]
4. *Think before you post*: There are no takebacks on the Internet. Every click, and post, and message can be cataloged, so be careful out there and remember, *the Internet is written in ink*!
5. *Opt out of mailing lists*: to decrease the number of third parties with your information, unsubscribe from mailing lists and try not to sign up for too many such services unless you're genuinely interested in the product or offering.

6. *Use a VPN*: A virtual private network (VPN) encrypts the data that you send and receive via the Internet.[103] It's like creating a secure tunnel of communication in a city of windows. There are lots of great options, some free and some with a cost.[104]
7. *Consider using an anonymous search engine*: Major browsers, like Google's Chrome, track your activity and resell it to marketers. As has been mentioned, even opting out is no guarantee. So, there are alternative search engines, such as DuckDuckGo, that say they value privacy, though even this service has faced allegations that it allows Microsoft to insert trackers.[105]
8. *Update your software*: Don't keep clicking "remind me later." Ideally, you should enable automatic updates so that they are installed while you sleep. These patches help make it harder for your system—and digital footprint—to be compromised.
9. *When in doubt, don't*: Be conscious of what you click on, especially in unsolicited emails and on the Web. When in doubt, double check, especially before initiating a wire transfer. The most common attacks use a method called "phishing," or a variant that specifically targets one potential victim, called "spearphishing." These typically take the form of email messages that appear to be sent by coworkers or supervisors asking for sensitive information.
10. *Use strong passwords and multi-factor authentication*: Passwords should be long, and complex. Keep them secret and change them often. Consider starting with a favorite sentence, and then just take the first letter of each word. Add numbers, punctuation, or symbols for complexity. Use a password manager if you have trouble remembering them all, or a biometric option, and be sure that any sensitive information is encrypted. And, whenever you're prompted, enable multi-factor authentication.[106]

None of these are panaceas; they all have their problems but, again, the more layers of Swiss cheese that we layer up, the harder it is for hackers to find a way through them all. Also, don't forget to check your credit report regularly for fraudulent activity (you might even consider freezing your credit until you need it).[107]

More generally, on the Metaverse, consider the three "Bes." First, be *aware* of who you friend, as they could be a vector to gather information for a future phishing attack. Second, be vigilant and organized, including by having data backups in case your system is compromised in a ransomware attack (ideally, this should include a copy of the data on site, and in the cloud). Third, be proactive. Don't wait for a worst-case identity theft or ransomware scenario to unfold. The Australian government was able to dramatically decrease successful penetrations by taking three basic steps: restricting which programs can run on government computers, keeping software updated regularly, and minimizing the number of people who have administrative control over networks and key machines. Remember, this stuff isn't rocket science, it's just computer science, and you don't have to be Neo to protect yourself in the Metaverse. Ultimately, though, to protect your identity and privacy more completely in the Metaverse, privacy and cybersecurity laws may need to be updated at the state, federal, and international levels, topics that are addressed in the next two chapters.

5

IDENTITY IN THE METAVERSE

I do think that a significant portion of the population of developed countries, and eventually all countries, will have AR experiences every day, almost like eating three meals a day. It will become that much a part of you.

–*Tim Cook, Apple CEO*[1]

Who are you in the Metaverse? Do you want a photorealistic version of yourself to act as your avatar? Or someone, or for that matter some*thing*, entirely different? What rights do you have to this persona, and should it operate across platforms? This chapter explores these issues—and more.

What does it mean "to be" in the Metaverse?

The Metaverse is not a physical location like Disneyland. Virtual reality headsets do not physically transport us into another location like the transporter from *Star Trek*. Nonetheless, we talk about "locations" in the Metaverse all the time. We talk about digital places where individuals can shop and experience virtual events.[2] In these locations we play, work, and chitchat. We do so by inhabiting our avatars, the 3D representations of ourselves through which we interact with the Metaverse.[3] Does this virtual presence translate into truly "being" in the Metaverse?

To begin answering this question, let us first compare avatars with something that may look quite similar: a video game character. Specifically, let us look at video games that most closely resemble the Metaverse: games in which players control a primary human(oid) character who serves as the game's protagonist. In such games, players are frequently asked to choose from a small number of predesigned character models which are clearly not meant to represent the player's identity. When playing Super Mario Odyssey, for example, we are not asked to personally identify with Mario when controlling his character model.[4] When playing Halo Infinite, we do not have to see ourselves in the Master Chief.[5] Even in a role-playing game like Final Fantasy XIV—where individuals are invited to spend hours "[t]weaking [their] character's height, eye color, hairstyles and color, beard options for male characters, bust sizes for female characters, scars, tattoos, and much more"[6]—players have a clear boundary between themselves and the characters they control. When gaming, we are asked to, and frequently want to, leave our true selves behind, and to adopt the body, identity, abilities, and goals of these fictional characters. However, they do not represent us.

In the Metaverse, however, avatars are indeed meant to represent us. They are meant to be a virtual version of ourselves. Unlike video game characters, avatars are not preexisting entities with their own looks, personalities, and behaviors. This distinction is evident in the way Metaverse companies discuss the very purpose of avatars. Meta's vision of the Metaverse, for example, is replete with identity-centric avatars. In a 2022 press release, Meta emphasized the relationship it envisioned between users and their avatars:

> Your avatar is a digital expression of your personality (or personalities). It can convey how much of an extrovert or introvert you are, your sense of humor and even your fashion sense. *It lets you be your authentic self*, and that in turn can help you connect more meaningfully with your friends, family, coworkers, or anyone else you meet on the road to the metaverse. We want to enable everyone to *present the best version of themselves*, which is why we've made it possible to

customize your avatar with a wide variety of free outfits and accessories. Feel like giving yourself blue hair today, trying out new makeup or sporting a suit for a professional look? No problem—you can customize your avatar anytime you want to suit your mood. In fact, with more than a quintillion combinations of free avatar options already available, your choices will almost certainly be uniquely yours.[7]

Even companies that want to avoid the term "avatar" nonetheless recreate the concept due to its status as a core infrastructural part of the Metaverse. When discussing its VR headset, the Apple Vision Pro, Apple avoided terms like avatar or Metaverse. Instead, Apple's communications focused on the more familiar-sounding productivity and entertainment features of this new "revolutionary spatial computer."[8] Despite this marketing tactic, the company spent a significant amount of resources in developing the technology for what Apple calls a "persona."[9] According to the company, a persona is a "a digital representation" of a user "which reflects face and hand movements in real time."[10] Through the use of an "advanced encoder-decoder neural network," the headset recreates "your face with a hyperrealistic" digital rendering.[11] In other words, your persona is your *avatar*.

For all intents and purposes, Meta and Apple seem to want this 3D rendered version of yourself to *be* you. To the extent this is true, these companies are not necessarily wrong. Your avatar is the one and only way through which you can express yourself in the Metaverse. It is the one and only way through which your ideas, thoughts, actions, and various other characteristics are present in the Metaverse. What is identity but a collection of ideas, thoughts, actions, and various other characteristics? In many ways, then, your avatar truly *is* you in the Metaverse. Your boss, your mother, your best friend, for example, will be looking at this 3D rendering as if it were you because, to them, it is you.

The relationship between physical self and digital self is a large part of what makes the Metaverse different from an immersive but fictional video game. Without this relationship, the Metaverse is little more than yet another imaginative digital space.

Companies, who stand to profit from the success of the Metaverse, then, are understandably doing everything in their technological power to enable avatars to truly carry our personal identities. Cynically speaking, it is only by achieving this goal, that companies will be able to translate the human yearning for "authentic self" expression into the sale of "digital outfits from Balenciaga, Prada, and Thom Browne."[12] Optimistically speaking, it is only by achieving this goal that we as individuals can experience the full benefits of the Metaverse.

As the examples above indicate, companies such as Apple and Meta are designing avatars to *be* a version of ourselves. By accomplishing this, they can dissolve the distinction between our physical and digital selves. Importantly, an avatar is not designed to be an intermediate interface mechanism through which we interact with the Metaverse; rather an avatar is meant to be a true version of ourselves. But what does this distinction even mean in practice?

Imagine that you are using a mouse to browse an ecommerce website. You control the mouse such that it scrolls through and moves around the website. Ultimately, you use the mouse to click "buy." Did the mouse—or the computer cursor—make the purchase? No. *You* made the purchase using a physical-to-virtual interface. An avatar is not meant to be like a mouse. Rather it is meant to be you, your hand, and your body.

Is this level of techno-psychological integration even possible? Would our minds truly accept that reality? Research on this topic has identified various aspects to consider when addressing such questions: (i) immersion, (ii) interaction, and (iii) embodiment.[13] While experts disagree on the nuances of each of these factors, the following short discussion provides an overview of the relevant research.

What is "immersion" in the Metaverse?

The word "immersion" is frequently used to describe any captivating activity. For example, you might say that you're immersed in a novel or in a jigsaw puzzle. In the context of the Metaverse, however, its meaning is more specific.[14] For the Metaverse, immersion refers to perceptual presence, "a psychological state characterized by perceiving oneself to be enveloped by, included in, and interacting with an environment that provides a continuous stream of stimuli and experiences."[15] Immersion is achieved when users "feel, act or react as if actually present" in the Metaverse.[16]

To succeed in establishing immersion, three technical conditions must be fulfilled: "sensory motor loop, statistical plausibility and behavior-response correlations."[17] In layman's terms, immersion requires an integration with your senses such that your responses to virtual stimuli achieve the desired result. For example, in an immersive scenario, when you sense a ball coming toward you, you desire to raise your arm to catch it, so your avatar does indeed raise its arm to catch the ball, and the ball is caught. Generally, to achieve statistical plausibility, true immersion also requires that the virtual world respect real-world physics (or a similar set of rules) such that we can come to anticipate causes and effects. Without these factors, it will be much harder to be immersed in the Metaverse.

What about "interactivity"?

Simply being immersed in a virtual environment is not enough to achieve the goal of *being* in the Metaverse. Imagine watching a nature documentary but in virtual reality. It is possible to be immersed in the visuals, sounds, and storytelling as a passive observer without truly being in the documentary. Such experiences do not transport you into the Metaverse; rather, they bring externally defined and controlled media to your senses. Once an environment is interactive, individual users gain a sense of agency that separates a true Metaverse from other virtual experiences.[18] Interactivity allows individuals, through their avatars, to go toward new experiences and senses. Through interactivity, digital media is no longer being served to a passive user; instead, the user becomes part of the ever-changing Metaverse.

What does "embodiment" in the Metaverse mean?

As noted above, an avatar is not designed to function like a computer mouse. The theoretical relationship between user and avatar is more complex and is defined through the concept of embodiment. Through embodiment, your avatar becomes "invisible" to you. To clarify what invisibility means in this context, let us look at the physical world. When taking a walk, do you notice and think about your legs moving? Your arms swinging? Your feet touching the ground at each step? Not usually. While walking, your body and its individual movements become invisible to your mind. When done successfully, walking just works. In the Metaverse, your avatar should become equally invisible to you. It should just work.[19] Counterintuitively, such invisibility also turns your avatar into a highly salient means of self-actualization: "Digital bodies tell the world something about your self. They are a public signal of who you are. They also shape and help make real how users internally experience their selves."[20] Embodiment is the last step in making your avatar a true digital version of your physical self.

With immersion, interactivity, and embodiment, we can truly *be* in the Metaverse. Our digital presence, through avatars in the Metaverse, may be just as real and meaningful as our physical presence in the physical world.

Can I bring all aspects of my personal identity into the Metaverse?

In the Metaverse, personal identity is manifested primarily through avatars, users' digital representations. Avatars can be designed to closely mirror a user's physical-world appearance and characteristics, or they can be constructed as a complete departure from one's physical identity. Meta's promotional materials, for example, note that the Metaverse allows you to "share *any version* of you . . . with almost unlimited ways to customize your avatar."[21] This flexibility allows users to experiment with and express aspects of their identities in ways that may be unusual, impossible, impractical, or even dangerous in the physical world.

Research has shown, for example, that "people with more attractive [avatars] report that their behavior online is more extroverted, loud and risk-taking" than their behavior in the physical world.[22] Such ability to experiment with various identities, appearances, and abilities is incredibly freeing but, as this section later explores, this flexibility may also be the cause of several types of social conflict in the Metaverse.

Before discussing the possible pitfalls, however, let us acknowledge that, unlike video games, the Metaverse is purposefully designed so that users can bring with them their true and complex individual identities. In this way, the Metaverse offers a space for individuals to explore and express aspects of their identities, like race, gender, and sexual orientation, which may be central aspects of their core selves. By facilitating this self-exploration, avatars can become "strongly interwoven into users' self-perception and virtual self-identity in the [M]etaverse."[23] In other words, avatars can truly be a means of virtual embodiment.

Without the constraints of the physical world, the Metaverse also enables the exploration of individual identities that would engender prejudice or even "criminal sanctions" in the physical world.[24] Imagine, for example, members of the LGBTQ+ community that live in jurisdictions in which their very existence is criminalized. These individuals may be physically and legally prohibited from expressing themselves and living their true lives. In the Metaverse, however, they may be able to choose gender-affirming avatars, enter into relationships, and create a social support network. Such freedom of expression is a way through which the Metaverse can enable a deeper understanding and acceptance of oneself and of others.

This freedom of identity inherent to the Metaverse also allows individuals to adopt personas that completely bypass culturally defined categories. A Meta community forum post, for example, noted that "[t]he best part of VR is escaping my boring human existence."[25] This individual went on to complain about situations in which realistic human avatars were the only options: "Why do I still have to look like a human? I can never even make the human avatar look like ME anyways, and it drives me crazy. Robots, Monsters, Animal-People, all would be great to have as avatar options."[26]

Nonhuman avatars aside, bringing personal characteristics into the Metaverse could promote a more inclusive virtual world by enabling the presence and representation of a wide array of individual identities. Such a digital environment could foster empathy and understanding among users from different backgrounds, who would potentially never interact in the physical world.

Unfortunately, the digital implementation of personal identities could also lead to the recreation of biases, stereotypes, and other forms of discrimination in the Metaverse. One possible area of concern is the way in which identities are digitally represented and perceived in virtual spaces. For example, if racial features are implemented through a pre-built options menu in an avatar-creation screen, many problems may emerge. This is exactly what happened in the 3D simulation game *The Sims 4*. Critics argued that "the game ha[d] a blind spot about race" and was "creatively paralyzed," unable to "conceiv[e] of [race] as anything other than a superficial customization."[27] Some players noted that "[c]ustom mods [were] necessary . . . to create good-looking Black Sims" not least because "more/better skin tones" were originally missing from the game.[28]

It is also important to note the relationship between anonymity and personal identity in the Metaverse. On the one hand, anonymity allows individuals to create avatars to avoid any identity-based prejudice or persecution they might experience in the physical world. This ability weighs against any requirement to link one's Metaverse avatar to a physical and legal identity. On the other hand, the ability to adopt an avatar of any race, gender, or any other individual characteristic opens the door for possible abuse. Are the following situations acceptable? Could they lead to social conflict in the Metaverse?

A woman uses a male-presenting avatar.
A man uses a female-presenting avatar.
An Afro-Latino individual chooses an Asian-presenting avatar.
An Anglo-American individual chooses an African-American-presenting avatar.

Although similar questions may be less consequential in the context of video games, empirical research has begun to uncover interesting patterns in user behavior. One study found that "29% of men prefer playing female characters" and "9% of women prefer playing male characters" in video games.[29] What does this mean for the Metaverse? It is possible that the gender and race of avatars may offer little information about the characteristics of the humans behind the curtain. Other research, however, notes that despite choosing female characters in video games, this 29% of men "may not necessarily seek to mask their offline gender [in game] when they use a female" game character,[30] possibly noting that this statistic does not translate directly to Metaverse avatars.

As the number and diversity of Metaverse users grow, it is crucial to ensure that all individuals are free and able to embody their personal identities. This task will take two forms, each with its own set of dangers. First, the Metaverse must facilitate the virtual representation of physical-world identities, which will enhance diversity and inclusivity. The implementation of these characteristics, however, must also avoid perpetuating stereotypes and inequalities. Second, the Metaverse must also offer opportunities to rethink and potentially transcend traditional notions of identity, shifting the focus of personhood from predetermined categories to more fluid and self-defined identities. This freedom, however, must not come at the cost of respecting the existence, autonomy, and dignity of cultures, ethnicities, genders, sexual orientations, and the many other dimensions of identity that are core to our human experience.

Does anonymity exist in the Metaverse? Should it?

As it currently exists, the Metaverse offers a certain level of anonymity. Users can create avatars that do not necessarily reflect their physical-world identities. Users can also connect to the Metaverse via VPNs or through other means of masking their physical locations. This anonymity can be partial or complete, allowing individuals to interact in a space that is as detached from their physical identity as they would like. Some users may opt for avatars that are true to their physical appearance using their legal names; others may choose fantastical or entirely different personas, as discussed earlier in this chapter. In this section, we will discuss the question of whether anonymity should exist in the Metaverse and dissect various ethical, legal, and social dimensions of this question.

Before diving into the Metaverse, however, let us first look at the effects of anonymity in the context of the Internet as a whole. Research has shown that anonymity is beneficial in several ways. First is the fact that anonymity enables "privacy," a foundational pillar of "psychological well-being."[31] The technical ability to hide one's identity online does not mean that one has to be completely anonymous at all times, rather "it merely involves one's ability to exert boundary control upon others' access to one's self."[32] In other words, anonymity protects personal autonomy on the Internet by granting individuals a choice over how much of their identity to reveal.

A similar, though distinct, benefit is that anonymity can "create a more equal playing field for communicators."[33] This means that "individuals who traditionally possess less power in society (e.g., women, minority group members, individuals with disabilities) should have increased power in an [anonymous] Internet environment" because "individuals are unable to project stereotypes on others and thus, [lack] expectations for behavior based on these stereotypes."[34] Anonymity is also helpful in preserving freedom of expression online. This freedom is useful for more than just voicing unpopular opinions. Rather, the freedom to communicate anonymously is helpful if not necessary in:

> (a) facilitating the flow of communication on public issues without killing the messenger (e.g., tiplines, whistleblowing, unsigned political communication, etc.); (b) obtaining sensitive information (such as in research); (c) focusing attention on message content rather than status of source; (d) encouraging reporting, sharing, etc. for stigmatized situations; (e) protecting one from subsequent contact (e.g., anonymous donors); (f) avoiding persecution and retaliation for one's beliefs; (g) encouraging risk-taking, innovation, and experimentation; and (h) enhancing play/recreational interaction.[35]

With these social and political benefits in mind, it is unsurprising that the right to communicate anonymously has been routinely protected under the First Amendment.[36] The Supreme Court of the United States, in a 1960s case about anonymous political pamphlets, noted that:

> Anonymous pamphlets, leaflets, brochures, and even books have played an important role in the progress of mankind. Persecuted groups and sects from time to time throughout history have been able to criticize oppressive practices and laws either anonymously or not at all . . . Before the Revolutionary War colonial patriots frequently had to conceal their authorship or distribution of literature that easily could have brought down on them prosecutions by English-controlled courts . . . Even the Federalist Papers, written in favor of the adoption of our Constitution, were published under fictitious names.[37]

Given these policy considerations (and constitutional protections),[38] it would be hard to imagine a Metaverse in which anonymity were banned completely. Nonetheless, online anonymity poses significant dangers to safety and accountability, and counterintuitively, may even undermine the free and open communication it seeks to enable.

Online anonymity can lead to irresponsible behavior, as individuals who believe they are untraceable may engage in harassment, cyberbullying, and other harmful activities. Anonymity may also empower individuals to engage in activities that are not only harmful but actually illegal, such as libel, slander, and defamation.[39] In such cases, First Amendment protections would be unavailable as a defense, but if technical anonymity truly hides the identity of the perpetrators, the law may be powerless.[40] Ensuring accountability in such a context may be extremely challenging, raising difficult questions about how to balance the freedom of anonymity with the need for security and civility.

Online anonymity may also harm the free marketplace of ideas. Without knowing speakers' identities, "[a]udiences have a harder time evaluating credibility, and there is more room for deception and frivolousness."[41] Online anonymity may also undermine "trust, because we know relatively little about [the] personal character" of the individuals with which we interact.[42]

A Metaverse without accountability and trust does not sound like an attractive destination, thus it is hard to imagine a successful Metaverse in which there is complete anonymity. We thus find ourselves in a situation where the Metaverse must both enable and limit anonymity. In such a case, Metaverse platforms (and the regulators that govern them) will need to decide how to thread this needle. The existence of anonymity in the Metaverse allows for a degree of freedom and exploration that is liberating and empowering. Without this freedom, it is unlikely that the Metaverse will grow into the immersive and free world it aims to be. However, Metaverse platforms must also provide robust mechanisms to ensure that this freedom does not lead to harm or lawlessness. The future of anonymity in the Metaverse should be a subject of ongoing dialogue among technologists, legal experts, ethicists, and the broader user community.

Will the Metaverse foster and fairly treat diverse participants?

Recognizing and respecting the diverse array of human differences and ensuring fairness are principles that have traditionally been most relevant to physical spaces. The role of these principles in the digital realm, especially in the Metaverse, is both significant and complex. This is especially the case because the physical boundaries and borders that kept individuals apart in the physical world are irrelevant in the Metaverse, enabling a new level of interaction between people of different cultures, identities, and experiences. It is important to note that the brief discussion presented here does not even attempt to adequately scratch the surface of any sufficiently comprehensive analysis of these issues in general, in the digital realm, or even specifically in the Metaverse. Nonetheless, as a crucial element of a successful Metaverse, these principles will be relevant to almost all other topics discussed in this book. Thus, the following paragraphs are an attempt to highlight a few of the many ways in which such considerations are especially relevant to the Metaverse.

Diversity, as a general term, can be defined as "the condition of having or being composed of differing elements."[43] Under this definition, diversity is an inherent aspect of human existence. Our species—with its eight billion individuals who live in different places, do different things, and think different thoughts—is a prime example of diversity. This reality, however, has not been just a neutral historical fact. Rather, this global state of diversity has been culturally, economically, and even medicinally beneficial throughout human history. When discussing the ancient Silk Road, for example, historian James A. Millward summarized that "humanity has thrived most when connected across its far-flung habitats by exchanges of goods, ideas, arts, and people themselves."[44] Given the potential benefits of exchanges across diverse groups, it is no wonder that companies, communities, and other institutions have invested in efforts to foster diversity. Any such benefits likely extend to the Metaverse, giving cause to ensure the digital environment is a diverse one. To that end, the Metaverse might have a leg up over physical institutions. As a digital public sphere, accessible from anywhere in the world, by anyone, at any time, the Metaverse is primed to host the true array of human diversity, whether based on race, ethnicity, background, ability, gender, sexual orientation, neurodiversity, philosophy, and beyond. To truly unlock the benefits of wide-ranging diversity, however, the Metaverse must have built-in tenets of fairness and accessibility.

In terms of access, the Metaverse is built for and available to individuals who: (i) can afford the technology necessary for virtual reality, (ii) have high levels of digital literacy, (iii) live in areas with high-speed broadband internet, and (iv) speak English or one of the small number of languages in which Metaverse content is made available. A small fraction of individuals in the world fit these requirements. For the rest of the world's population, the Metaverse and all its benefits remains inaccessible. By fostering fair treatment and accessibility, this "digital divide" will likely start to close and thus increase technological participation.[45] Also by fostering fairness and accessibility, the goal of building the Metaverse into a vibrant and diverse digital environment will be more attainable.

Fairness and accessibility also pertain to how welcomed, respected, and valued individuals feel in any particular environment or situation. A fair and accessible environment is one in which diverse voices are heard, considered, and respected in decision-making processes. In the context of the Metaverse, the benefits of such an environment exist both within the digital realm and among the tech companies involved in the development of the Metaverse. As a product of human design, the Metaverse is subject to all the biases and perspectives of its creators. Therefore, developers would likely do well to proactively create inclusive environments in which everyone feels encouraged to voice their ideas and concerns. Otherwise, the Metaverse runs the risk of digitally reconstructing or even exacerbating unaddressed injustices that exist in the physical world, thus falling short of its full potential.

Unfortunately, promotion of the principles of fairness, accessibility, and the fostering of diverse participation in the Metaverse will not be straightforward. The anonymity and flexibility of identity that the digital world offers adds a layer of complexity that comparable efforts in the physical world have not had to deal with. Unlike in the physical world, Metaverse users can choose to digitally adopt (or not) each personal characteristic of their physical identities. This power of selective anonymity is neither inherently good nor inherently bad.[46] User A might opt for an avatar that represents their physical race, ethnicity, and gender. User B might opt for an avatar that does not align with their physical race, ethnicity, or gender. User C might opt for a nonhuman avatar all together. This freedom may be a crucial element of the Metaverse but will also present complex challenges.

As the Metaverse grows and evolves, it will provide us with an opportunity to redefine norms and create wider-ranging digital experiences with enhanced participation. Taking full advantage of this opportunity, however, will require a continuous conscious effort from developers, regulators, and users alike.

6
PRIVACY IN THE METAVERSE

Privacy and safety need to be built into the metaverse from day one.

~Meta CEO Mark Zuckerberg[1]

Largely through digitization, consumer behavior (disclosure of personally identifiable information, shopping and purchase activity, information consumption, etc.) leaves a significant data trail, often at the disposal of various firms with which consumers engage. The now-old adage of "If you're not paying for the product, then you are the product," often comes to mind; here "you" means "data about you." The idea that such data have value is not new. As a very simple example, a basic listing—even on a computer printout (gasp!)—of names and addresses of households in the top 10% income bracket would be valuable to firms such as Rolex and Mercedes dating back many decades. The difference between then and now is that data now are collected, stored, and easily transmitted at volumes orders of magnitude higher than before.[2]

Immense increases in computing power and storage along with capabilities of communications technologies have flipped the cost effectiveness of data uses; many of the ways data are being utilized and traded today would not have been cost effective—or perhaps even possible—even just 10 years ago (and almost certainly not cost effective 20 or 30 years ago). Consequently, any potential risks of such data utilization and trading would not be a concern to their subject(s) in prior decades. Further, with relatively rare exceptions in the United States (e.g., personal health and financial information), until recently, firms have had virtual free reign on what they do with the data. In essence, firms have, by default, been given the property rights over the data—the data were theirs to do with as they please.

With the emergence of the Metaverse, the exchange of data and, consequently, concerns about privacy are set to explode. In this chapter, we discuss online data privacy as a general matter, data exchange and privacy in the Metaverse, the pros and cons of—and limitations on—privacy in the Metaverse, and Metaverse-relevant private and public data privacy protections.

What do we mean by "online privacy" in the Metaverse?

Understanding the type of data that typically gets exchanged online is crucial toward defining privacy in the twenty-first century, particularly in the Metaverse. Online activity generally involves the transfer of all kinds of personal data. Consider a few examples. To purchase a good and have it delivered to their home, customers will have to provide their home address. To purchase a flight, customers must provide their date of birth in addition to their name, email address and credit card information. Even to create an email address, customers generally must provide, among other things, their name, date of birth, and phone number. Full interaction with the online world effectively demands giving various product and service providers at least partial access to your personal information.

Data are also shared in the real/physical (offline) world, but often to a lesser extent. Walking into a bank branch and opening an account generally requires the teller to collect some of the applicant's personal details. Even if seeking employment offline, the employer generally will request a significant amount of the potential employee's personal information.

While data exchange is typical for both online and offline interactions, the *digital* nature of online collection, storage, and transfer of data sets it apart. To see this, consider the process of opening a bank account online and offline. Online, customers are already connecting to the bank over the Internet, and all information entered is immediately digital, allowing it to be immediately stored, processed, and transferred as the bank wishes. In addition, the bank easily can collect additional information not directly provided by customers, such as the time of day they began opening the account, the banking products they viewed while on the website, etc. Offline, the process of data collection is more cumbersome, involving the completion of paper forms or the oral communication of information to a teller in person or over the telephone. Collection of additional data not directly provided by customers offline is more difficult or even impossible.

The expansive amount of data that can and does get exchanged online can be sorted into some broad categories. The first and most basic category is personally identifiable information (PII). In short, PII is any information that permits the identity of an individual to whom the information applies to be reasonably inferred by either direct or indirect means.[3] For example, PII can identify an individual *directly* if it includes name and address; PII can identify an individual *indirectly* if it includes a combination of sex, race, date of birth, and geographic indicator (e.g., Northeast United States). These data are usually used for the fulfillment of a transaction, such as a product purchase, or to process an application. Additionally, they are also used by advertisers to "classify users into different demographics based on relevant parameters" and, thus, understand who interacts with their ads.[4]

Beyond PII, usage data such as "interactions with a business' website, text messages, emails, paid ads, and other online activities," are also recorded to build accurate consumer profiles. These data are also used to predict the type of content that best resonates with a consumer.[5]

Another type of data commonly collected online is behavioral data. These data include "purchase histories, repeated actions, time spent, movement and navigation on [a] platforms, and other types of qualitative data."[6] This type of data offers platforms and sellers information on consumers' tastes and helps marketing teams offer customers better fitted products.

In Table 6.1, we present a more comprehensive list of the types of data commonly collected online. While these are the most common types of data getting exchanged online, as we explain in detail below, the scope of the data that will be gathered and exchanged in the Metaverse will most likely expand this list.

In the broadest sense of the word, "privacy is the right to be let alone, or freedom from interference or intrusion," while "information privacy is the right to have some control over how your personal information is collected and used."[7] Put another way, information privacy is "the interest an individual has in controlling, or at least significantly influencing,

Table 6.1 Types of Data Exchanged Online

Data Category	Examples
Personal Information	Name, address, phone number, email address
Financial Data	Credit card information, bank account details, tax records
Health Data	Medical records, prescription history, fitness and activity data
Location Data	GPS coordinates, location history
Communication Data	Chat history, emails, voice calls, video calls
Search and Browsing Data	Search history, websites visited
Social Media Data	User profiles, posts and comments, likes and shares
E-commerce Data	Product reviews, shopping cart contents, purchase history
Government and Legal Data	Court records, criminal history, legal contracts
Educational Data	Student records, academic transcripts, online course progress
Environmental Data	Temperature, air quality
Media and Entertainment Data	Music and movie playlist, streaming history, video game activity
Research and Scientific Data	Scientific data sets, research papers, experiment results
Business Data	Financial reports, inventory information, supply-chain data

the handling of data about themselves."[8] With the development of the Metaverse, the term information privacy has become particularly prominent, and alternative definitions have emerged. Nonetheless, while "there are many definitions for information privacy, . . . there is little variance in the elements of the definitions, which typically include some form of controls over the potential secondary uses of one's personal information."[9]

One way to achieve information privacy would be to completely withhold personal information. However, given the number of interactions and transactions an individual typically undertakes in a day, such an approach is virtually impossible. Standard transactions such as the opening of a bank account or issuing of a driver's license involve the exchange of personal information, whether online or offline. Recall, though, that information privacy is breached only when data is not used for the purposes for which it was handed over to a third party in the first instance. Therefore, full information privacy would be achieved if this criterion were met, even if substantial amounts of data are exchanged. For example, if customer data provided to a bank to open a bank account were used exclusively by the bank to service that account in ways intended by the customer, the customer has retained full information privacy in this instance.

While information privacy is technically possible for virtually any online or offline interaction, there are several ways that individuals can and often do lose information privacy. First, the entity, say a firm, collecting the data may use the data itself in ways for which it did not receive consent. For example, it may use the data to recommend its own products. In addition, the firm may share the data with third parties in exchange for financial or other forms of payment; for some firms, the selling of their customers' data constitutes their main source of revenue.

The selling of data to third parties raises additional privacy issues beyond ones that may arise when examining a single customer/firm interaction. Specifically, third parties may obtain different data sets pertaining to the same individual from different sources, where each data set on its own would not be considered PII; however, the *combination* of the different data sets may be such that "data scientists and other savvy investigators can combine de-identified data in a way that makes cross-references and re-identification possible."[10] Put another way, it may be that the combination of data sets that, by themselves, *are not* PII, results in a new data set that *is* PII. Hence, on each occasion of data exchange, individuals may (correctly) believe they are not providing PII, but with the sale of the separate data sets to third parties, the combination effectively becomes PII in the hands of firms with whom the individuals never even engaged.

Figure 6.1 is an illustrative depiction of the data flow from an individual's online book purchase. The solid line signifies data flows to which users willingly consent, while the dotted line indicates data flows to which users may not necessarily consent. For instance, when an individual searches online and buys books, their search data is collected by the search engine they use. In addition, financial data is transferred to the merchant to complete the transaction. Similarly, the bookstore must provide the courier company with the customer's address, name, and phone number for successful package delivery. Meanwhile, some data may be shared with third parties without the user's explicit consent, as represented by dotted lines. When an individual conducts online book searches, their search history and viewing behavior can be shared with advertising agencies, enabling them to deliver personalized ads to the user. The online bookstore can also use the data to recommend additional products to their customer.

Individuals can also lose information privacy via data breaches, where third parties gain unauthorized access to confidential information. While individuals may have some sense—or even be explicitly informed—about how their data are being used when deliberately shared with third parties, typically there are no actual or implied restrictions on how the data are used once they've been breached.

There are several measures you can take to protect your information privacy, as was introduced in Chapter 4. As an initial step, you can make information privacy one of the criteria you use in choosing the firms with whom you interact. You can do this by consulting the rankings and commentary provided by various organizations concerning companies' privacy protections and policies. In addition, you can take measures that constrain the amount of data firms are able to collect from you, as well as the vulnerability of your data to unauthorized access. For example,

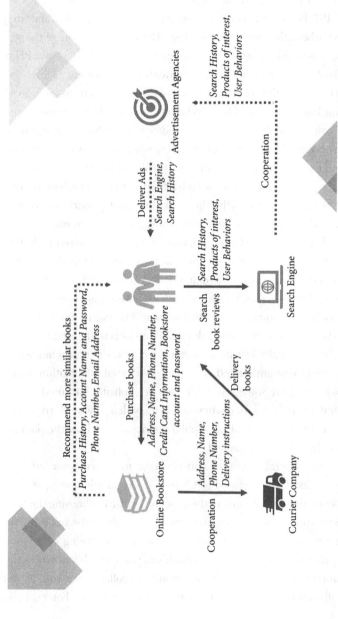

Figure 6.1 Data Flow from an Individual Book Purchase

you can utilize virtual private networks (VPNs), which encrypt customers' Internet connections and hide their IP addresses, thus enhancing their online privacy. You can also take various measures to protect against unauthorized access, such as utilizing the following:

1. *Proper password management tools,* which help generate, store, and autofill strong, unique passwords for all online accounts.
2. *Antivirus software,* which helps protect customers' devices from malware.
3. *Regular backup of personal data* on cloud-based platforms, which helps customers store their data securely.
4. *Data backup services,* which use end-to-end encrypted messaging apps for private communication.
5. *Secure messaging apps,* which prevent unauthorized access to your conversations.
6. *Multi-factor authentication (MFA),* which adds an extra layer of security to customers' online accounts.

Firms also take measures to protect their customers' information privacy, largely focusing on prevention of unauthorized access. Firms attempt to protect data from unauthorized access in various ways, including the following:

1. *Data loss prevention (DLP) tools,* which can help monitor and prevent the unauthorized transfer or sharing of sensitive data.
2. *Endpoint security solutions,* which provide comprehensive endpoint security, protecting devices from malware, ransomware, and other threats.
3. *Security information and event management (SIEM) systems,* which help organizations collect, analyze, and respond to security events.
4. *Encryption software,* which offers data encryption for companies' devices and files.

5. *Security awareness and training platforms,* which, along with corporate internal training, provide cybersecurity training for employees.
6. *Secure email gateway (SEG) solutions,* which provide advanced email protection.
7. *Data backup and recovery services.*
8. *Remote desktop solutions.*
9. *Secure communication and collaboration apps.*
10. *Cloud identify and access management (IAM) solutions.*

Firms also can protect users' privacy by voluntarily refraining from collecting information from their customers. There are several reasons why firms may be incentivized to collect less customer data. First, it may enhance the value of their product or service to their customers who value privacy and hence increase their willingness-to-pay. Second, collecting fewer data can reduce risk exposure from a data breach. And third, it can reduce the risk of reputational costs from customer perception of deception (if they didn't know certain data were being collected) or overreach ("the firm knows too much about me").

Lastly, firms that control platforms (e.g., the Apple App Store) can set rules for their platform meant to safeguard individuals' data privacy. Apple recently took this approach for its launch of iOS 14.5 in May 2021, during which it introduced the App Tracking Transparency (ATT) initiative[11], which mandates that "any developer sharing user data with a third party must obtain explicit permission to do so."[12] Users are presented with options to either consent or decline the sharing of their personal data.

Is privacy in the Metaverse any different than other online platforms or digital spaces?

The advent of cyberspace has catalyzed an explosion in data collection. For example, Instagram or an online clothes store can know how long you stare at an ad for a t-shirt on their site. This can be valuable information for the company looking to sell you its product or service. In the physical world, it is generally impractical—or at least typically less practical—for a retailer to monitor and record how long you are processing the information in front of you (e.g., the shop assistant could creepily monitor and use a timer or video surveillance). Moreover, a customer generally does not provide any sort of personal details upon entering a physical world location.

User data will almost certainly remain a sought-after item in the Metaverse, and data collection in the Metaverse will be able to go even further than the pre-Metaverse online world. As Mastercard's chief privacy officer recently put it in a blog post: "The Metaverse will be data collection on steroids."[13] In fact, Metaverse data collection will be able to nest (replicate and surpass) the data from pre-Metaverse online activity. Put another way, data in the Metaverse compared to the pre-Metaverse online world is "all that and then some." The expansion of data in the Metaverse generally stems from its immersiveness—the link it generates between the digital and physical worlds. Hence the Metaverse, which to a certain extent is a combination between the current online world and the physical world, will be able to broaden the scope of the data being collected and fill in some of the data gaps in the current offline and online worlds.

Consider some examples. In the Metaverse, you could be observed in different social environments, and therefore data could be collected on your behavior in various social circumstances, gaming activities, work environments, etc., including physical gestures. Moreover, like current online data, Metaverse data may consist of the amount of time you look at a piece of clothing before purchasing it. However, unlike current online data, it would be collected by observing your actual eye activity, in contrast to just observing how long a webpage is visible on your monitor. Even more, "the Metaverse could record how much blood is flowing to particular regions in users' brains," and eye-tracking data, such as "gaze direction and pupil reactivity, [which] may implicitly contain information about a user's biometric identity, gender, age, personality traits, drug consumption habits, fears, interests," and others. In other words, the Metaverse could create a user profile "that details subconscious states, mental conditions, or health issues."[14] As a result, the Metaverse, because it is seemingly recreating the real world, will be able to gather many of the data points that the real world was unable, or found impractical, to collect.

Table 6.2 highlights additional categories and examples of data that can be collected in the Metaverse. As it notes, Metaverse data contain and expand upon data types collected in the pre-Metaverse online world, listed in Table 6.1. Through the Metaverse, all these existing data sets could be layered with behavioral data, spatial data, and biometric data to truly create a 360-degree individual profile. Note that, while some of this type of data collection, particularly biometric, is already happening via facial recognition and fingerprint technology included in many mobile devices, these features are currently on an opt-in basis. In contrast, many AR and VR technologies *require* the tracking of bodily movements—such as eye position—as part of their general use.

Just as you can take measures to protect your information privacy in the pre-Metaverse online world (see the previous question), you will be able to take at least some of these protective measures in the Metaverse as well. For example, given Metaverse dependence on the Internet, you will be able to utilize VPNs to encrypt your connections. You will also be able to take various measures to protect against unauthorized access of your Metaverse account(s), just as you can for current online accounts (e.g., email).

There will also be Metaverse-specific protections that you can use to protect your information privacy, some already in existence. For example, researchers have already "developed a new tool called

Table 6.2 Types of Data That Can Be Collected in the Metaverse

All Data Listed in Table 6.1, Plus...

Data Category	Examples
Behavioral Data	Biological clock, physical movements when interacting with other people or digital entities
Spatial Data	3D environment data, geospatial data
Biometric Data	Facial recognition, voice, heart rate, blood pressure

MetaGuard, . . . which uses a technique called differential privacy to add noise to certain tracking measurements, making them less accurate and unable to identify users without impacting the user experience." This tool "operates similarly to 'incognito mode' in browsers and can be toggled on and off by users depending on the level of trust they have in the environment they are in."[15]

In principle, you will wield some control of your information privacy via your choices of firms you engage with for Metaverse activity; if you wish, you can steer your business to Metaverse firms that provide greater protections. However, the extent to which such choices will exist—and hence you will be able to base your choice on your understanding of firms' varying privacy protections—will depend on Metaverse economics and, among other things, the corresponding number of competing firms (see Chapter 2).

As an avatar in the Metaverse, you may have additional avenues toward protecting your information privacy, as described by the European Data Protection Supervisor:

> Depending on a particular case, certain Metaverse platforms could allow individuals to create avatars with entirely fictional characters that do not resemble the physical appearance or include any related information with the real person; or to create any other elements and objects relating to them having features different from the corresponding objects in reality, insofar as this might be considered fair and without negative implications for others.[16]

As a result, "this could be used to enhance anonymity towards the other users/vendors within the entire interacting process in the platform."[17]

Like cyberspace, firms will be able to use various tools to prevent unauthorized access (see previous question for examples); they may also find it in their interest to willingly forgo collecting some data to accommodate consumer preferences, reduce data breach risk, and/or reduce reputation risk. The incentives to show restraint may be heightened in the Metaverse, at least in its early development, to avoid any negative publicity on privacy and reduce the risk of falling behind in a competition for network size. It is important to keep in mind that "the Metaverse is a concept under development and the design and configuration of the technology is still not specified."[18] Therefore, "eventual privacy features could be implemented in the Metaverse to obtain an enhanced level of privacy."[19] Privacy in the Metaverse will be the result of a symbiotic relationship between dynamic firm capabilities and dynamic individual privacy preferences and corresponding behaviors. As consumers become more aware of the many ways their data can be collected and utilized, they may develop an increasing aversion to such practices. Ideas that conceptualize data use could also influence preferences to the extent that they strike a nerve with the public. For example, author Shoshana Zuboff describes an "economic system built on the secret extraction and manipulation of human data," labeled "surveillance capitalism."[20] If this framing takes root on a broad scale, individuals may become more demanding of privacy protections online, including in the Metaverse.

Is privacy desirable in the Metaverse?

In general, the collection and utilization of customers' data can result in a range of tangible benefits for those customers. The most commonly cited—and scrutinized—benefit comes from targeted advertising. When firms have information about you, they can try to ensure the offers and services marketed to you are relevant and personalized.

Targeted ads can yield several different types of benefits to consumers. First, they can help you to find better matches for your interests. For example, if you are in the market for a new car, and have no interest in buying new furniture, you are plausibly better off receiving a targeted stream of car ads rather than a random mix of car and furniture ads. If firms have information that reveals your car-vs.-furniture preference (e.g., from your web browsing behavior), they can deploy the ostensibly more beneficial ad stream. Targeted ads can also help you discover new products that you prefer and otherwise would not have discovered. For example, if advertisers can ascertain that you are into home fitness, they can send you targeted ads that may alert you to new fitness equipment that suits your tastes and that you hadn't known about. Hence, this facilitated discovery could lead to you being better off by buying the targeted equipment rather than an alternative you didn't like quite as much. Note that, along with ads targeted to your interests, you will also see better prices; that is, a targeted ad may alert you to a product you want to buy that is being sold at a lower price, thus saving you money and making you better off.

More broadly, targeted ads can help reduce the costs associated with finding the products and services consumers most prefer, known as search costs.[21] These costs can include the time and effort to find and assess products on the market by reviewing websites or visiting physical stores. If an ad is well targeted (i.e., concerns a good match for the recipient), you may avoid some or all the search costs you would have incurred to identify that good match. Well-targeted ads can also reduce the disutility from irrelevant ads; in the extreme, receiving randomly generated ads often will seem like a waste of your time, as many ads will concern products you don't care about. In contrast, you may find time spent viewing well-targeted ads to be useful if it leads to the purchase of well-matched products and services.

Firms can also use data to provide helpful recommendations, customization, and simple conveniences. If through data collection and analysis, a firm understands your preferences—say for movies—it can make recommendations for future viewing that cater to your tastes. For example, Netflix and YouTube collect data on the movies and shows you watch, as well as your reactions to them (e.g., star ratings for Netflix or thumbs up for YouTube). Based on such information, and like targeted ads, they can reduce your search costs for other movies, shows, and videos you would enjoy by making accurate recommendations. Similarly, through data, firms can customize products and services in ways that accommodate your preferences. For example, by sharing pictures of your feet, firms can customize their shoe offerings to your specific foot dimensions and contours.[22] As for convenience, the simple collection of financials and login data can be used to save time by automatically filling in fields required for regular online activities, such as making a purchase (credit card information) or logging in to various accounts (username and password). Data can also be used for public services and research that can benefit consumers and society at large. By just knowing your location, local authorities can send you specific messages to help you avoid incoming, dangerous weather. Firms can also use data—particularly through controlled experiments—to conduct research and gain deeper insights about their customers and possibly broader groups of individuals, which can lead to better product design at least by that firm and possibly industry-wide.

Data collection and utilization in the Metaverse generally can provide these same benefits but often to a greater degree, as well as additional benefits. Behavioral, spatial, and biometric data, with examples in Table 6.2, can allow for even better targeted advertising. For example, firms can utilize data on your spatial surroundings and (cramped) movements within them to send an ad for a space-efficient loveseat sofa to replace your dated, full-sized couch. These types of additional data can also help firms more deeply understand your characteristics and, as a result, effectively predict your preferences in various domains, such as the type of vehicle you are interested in or the kind of food you like to eat. Detailed data on your biometrics may also be utilized to predict your health status, providing opportunities for firms in the Metaverse to make useful healthcare recommendations and/or deliver tailored healthcare advertisements. Related is biometrically inferred data (BID). These data sets come from information inferred from behavioral, physical, psychological, and other nonverbal communication methods. Together with AI-based predictive modeling techniques described in Chapter 8, BID presents a new realm of possibilities for targeting audiences.

The new types of data collected in the Metaverse also allow for more customized products and services that can benefit consumers. For example, data collection and utilization via VR and AR technology can be used by psychologists and psychiatrists to "personalize environments for individual patients in aversion therapy." Patients can be immersed into a controlled environment within which the doctors can monitor them closely. Moreover, Metaverse data, combined with AR/VR technologies, can even enable remote medical treatment. Patients and physicians can interact virtually in real time and perform some basic physical examinations with VR/AR, like "distant body observation, touch, auscultation, and vital sign collection."[23] Metaverse data and technology will be able to assist doctors in complex surgical procedures, enhancing "precision and accuracy."[24] Relatedly, data from the Metaverse have the potential to facilitate early diagnosis and treatments for various diseases. Convenience and financial security also can be enhanced in the Metaverse; for example, payments, particularly in high amounts, could become more secure than ever if biometric data are linked to bank accounts, and payments are processed by scanning fingerprints or eye retinas or through vocal recognition.

Given all of these (and likely other) potential benefits, whether privacy is desirable in the Metaverse effectively comes down to personal preferences and how you weigh these benefits against your preference for privacy. A preference for privacy can be driven by various factors, including the risk that some data usage poses to the data subject(s). Risks include identity theft, doxing, and revelation of sensitive information, among other issues. Some people also may experience disutility due to the collection/utilization of their data, per se (discomfort, creepiness).

Simply put, if you find little value in the benefits of data collection and utilization in the Metaverse and have strong privacy preferences for the aforementioned reasons (and/or others), you will find high levels of privacy desirable in the Metaverse—and vice versa. While simple in concept, this can be a difficult exercise in practice, as both the personal benefits and costs can be difficult to quantify, although recent research is attempting this at a broader scale.[25] Preferences can also change over time. For example, research in the early 2000s showed people having a relatively high preference for location privacy (i.e., ensuring companies don't know where you are located at a given point in time). However, more recent research has shown a much lower preference for location privacy.[26] A plausible reason is that the benefits of sharing location information have gone up in the past 20 years (e.g., real-time maps showing nearby amenities, current traffic reports, ride sharing, etc).

Thus far, we have only considered the desirability of privacy at the individual level. However, since data sharing often has what economists call externalities—as applied here, you sharing your data can impact others and vice versa—there can arise what's known as a "prisoner's dilemma." As applied to data sharing and privacy, some people may find it individually undesirable to share their data, but everyone would be better off if everyone shared. For instance, consider health records: Suppose your doctor finds an abnormality and is trying to diagnose your condition. Table 6.3 provides a simplified summary of how the prisoner's dilemma may play out. In the table, we see that if everyone else is sharing their data, your doctor can properly diagnose your condition regardless of whether you share yours. On the flip side, if no one is sharing their data, your doctor cannot make a proper diagnosis, again regardless of whether you share yours.[27] So, if you place value on your data privacy, you rationally would not share; this is because you get the same diagnosis outcome as you would get if you didn't share (regardless of what everyone else is doing), with the added benefit of retaining your privacy.; If everyone makes that same calculation, you end up with no data sharing and no diagnosis. However, if everyone committed to sharing, the diagnosis can be made; and if people generally value the ability to make the correct diagnosis more than they value privacy for these data, everyone is better off.

Table 6.3 A Data-Sharing Prisoner's Dilemma

	Others Share Their Data	Others Don't Share Their Data
You Share Your Data	Proper Diagnosis	Improper Diagnosis
You Don't Share Your Data	Proper Diagnosis + Privacy	Improper Diagnosis + Privacy

The existence of externalities doesn't just complicate the answer to whether privacy is desirable in the Metaverse, they also lead to our next question...

Is privacy even possible in the Metaverse?

The existence of externalities from data sharing may place significant limitations on the level of privacy that is possible, no matter how careful you are with your own data. To see this, it is helpful to flesh out the notion of data externalities a bit further, starting with a more formal definition. Data externalities refer to "the phenomenon that data of some consumers reveal information about others."[28] As a simple example, suppose you regularly attended a virtual space in the Metaverse, and all the other attendees voluntarily revealed that they have children. Then, even if you've never disclosed this information, firms that know the parenting status of all the other attendees may reasonably deduce that you have children as well (perhaps it's a virtual location for PTO meetings).

When such externalities exist, even if you have strong privacy preferences and consequently severely limit your data sharing, it may have little impact if others don't share your privacy preferences and data-sharing practices. This issue has already been highlighted in recent research for online data.[29] The key insight is that, given others are sharing their data, and hence your ability to protect your privacy is limited, you have reason to be more willing to share your data as well—this is because the privacy cost from sharing your data is lower (firms will know about you from other peoples' data) but you still get the same benefits from sharing.

It is important to note that the externalities from data sharing can be both positive and negative. In the medical diagnosis example captured in Table 6.3, the externalities are positive—sharing of data by others can benefit you by helping you get a better diagnosis of your medical condition. There are various other ways that firms' access to other people's data can benefit you. Recommendations are a good example of this. As we noted earlier in this chapter, good recommendations can benefit you by reducing your search costs and helping you find better matches. These recommendations are based not just on information firms gather on you but also on data contributed by other consumers. A quintessential instance is the "Frequently bought together" section of Amazon.[30] Figure 6.2 shows a recommended bundle of purchases for consumers purchasing a vacuum.

As another example, online maps rely on user-generated data to provide real-time traffic information. Google Maps collects data from its users to "identify areas in which many users are driving but not moving,"[31] represented as yellow or red in the map. Similarly, users can self-report hazards, car accidents, or speed checks along the routes. If enough reports are made, they will be shown on the map to all users, which will help people avoid traffic dangers or unnecessary slowdowns.[32]

Data externalities can also be negative, in that they can make you worse off. The simplest examples involve various personal attributes you may not want publicly known for various reasons. These include our prior example involving your parental status, as well as other attributes such as race, religion, political affiliation, and sexuality.[33]

In the realm of the Metaverse, a platform that enables rapid and wide-ranging data collection, both positive and negative data externalities will almost certainly increase. For example, online shopping websites can customize and deliver frequently bought together products in AR/VR with 3D previews, enhancing the realism of the shopping experience. On the flip side, biometric and spatial/movement data provide additional dimensions via which firms can accurately predict your attributes (using others' data) despite your choosing not to reveal them.

Figure 6.2 Vacuum Products Frequently Bought Together on Amazon

What will be the role of regulation in protecting privacy in the Metaverse?

Recall that firms and individuals can and likely will take actions to protect their privacy in the Metaverse. However, there are also many data privacy regulations that are meant to protect consumers in the current online world and likely will impact the Metaverse to come. Over 130 countries already have implemented some form of legislation "to secure the protection of data and privacy."[34] Perhaps unsurprisingly, there is ample international variation in the range and depth of such legislation, consistent with recent research that shows varying privacy preferences across countries.[35] Nonetheless, data storage laws and regulations usually require data holders to keep data "securely and protected against unauthorized or unlawful processing, loss, theft, destruction, or damage."[36]

The most widely hailed and critiqued of these is the EU's General Data Protection Regulation (GDPR), which came into force in 2018. In short, GDPR requires companies to ask for some permissions to share data and gives individuals rights to access, delete, or control the use of that data.[37] In other words, it puts consumers in the driver's seat to determine what happens with their data. Violating the laws can result in substantial fines. For instance, recently Amazon was fined €756 million, and Instagram was penalized $403 million for violating GDPR.[38]

In contrast to the EU, "the United States doesn't have a singular law that covers the privacy of all types of data. Instead, it has a mix of laws that go by acronyms like HIPAA, FCRA, FERPA, GLBA, ECPA COPPA and VPPA."[39] This alphabet soup may be understood as follows:

- HIPPA—Health Insurance Portability and Accountability Act of 1996—required the creation of national standards to protect sensitive patient health information from being disclosed without the patient's consent or knowledge.
- FCRA—Fair Credit Reporting Act—federal legislation enacted to promote the accuracy, fairness, and privacy of

consumer information contained in the data gathered by consumer reporting agencies.
- FERPA—Family Educational Rights and Privacy Act—governs access to educational information and records by public entities such as potential employers, publicly funded educational institutions, and foreign governments.
- GLBA—Gramm Leach Billey Act—requires companies that offer consumers financial products or services like loans, financial or investment advice, or insurance, to explain their information-sharing practices to their customers and to safeguard sensitive data.
- ECPA—Electronic Communications Privacy Act of 1986—protects wire, oral, and electronic communications while those communications are being made, are in transit, and when they are stored on computers.
- COPPA—Children's Online Privacy Protection Act—aims to protect the privacy of children under the age of 13 by requesting parental consent for the collection or use of any personal information of the users.
- VPPA—Video Privacy Protection Act—makes it unlawful for a "video tape service provider" to knowingly disclose, to any person, "personally identifiable information" concerning any "consumer" of such provider without their consent and with a few exceptions.

Despite these existing laws in the United States, "(t)he data collected by the vast majority of products people use every day isn't regulated. Since there are no federal privacy laws regulating many companies, they're pretty much free to do what they want with the data, unless a state has its own data privacy law. . ."[40] As of this writing, thirteen now have comprehensive data privacy laws on the books that apply across the board, including to personal devices. States as diverse as California, Indiana, Utah, and New Hampshire have all acted laws in recent years, with fourteen more states currently considering such bills. This activity suggests that, in the future, the United States may slowly shift from the bottom up to a similar type of data privacy protection at a national level. As we've noted, the scope of the data that could be collected in the Metaverse extends well beyond what is currently collected online. However, it is yet unclear "how the Metaverse plans to use sensitive data outside of the immersive experience" and therefore, there is still a lot of uncertainty around which data privacy regulations would be applicable to the Metaverse.[41] It is also unclear whether current regulations have "what it takes" to protect these additional, and potentially sensitive, data points. GDPR is known for being one of the most stringent and toughest regulations aimed at protecting an individual's data. "Because Facebook [now Meta] decided to focus its Metaverse development in Europe, [it is suspected that] the GDPR will have a central role in regulating the Metaverse."[42] However, some argue that not even "GDPR is currently . . . equipped to protect Metaverse users from data misuse as the Metaverse will spur thorny issues as it gains user traction."[43] A recent measurement showed that "in 2020 each Internet user created 1.7 megabytes of data every second he or she was online." The Metaverse is expected "to increase data usage of each Internet user by twenty times in the next ten years."[44] Moreover, most likely, "multi-sensory experiences in the Metaverse will expand the scope of data privacy beyond the normal data points to include emotional, biometric, and physiological data, meaning users will be monitored at an almost forensic level. This will make enforcement of data privacy regulation far more challenging."[45]

Despite such challenges, some important privacy issues likely will become more pressing as the Metaverse becomes more prominent. The first concerns property rights. Data externalities don't just come from the fact that others' data-sharing decisions affect you; they also come from the fact that firms' decisions about how to handle and utilize your data impact your data privacy. Consequently, firms may make decisions whose benefits don't fully outweigh their privacy costs (since the firms don't experience privacy costs). A classic solution to externality problems like this one is to allocate property rights, and such a solution for data privacy has been offered in the form of consumer data ownership.[46] Such an approach could be effective if it leads to firms fully accounting for consumers' privacy preferences, (e.g., by needing to compensate consumers when firm usage and/or sharing comes at the expense of consumer privacy).

However, granting consumers ownership of their data doesn't come without challenges. For example, the effective information the data carry often is a combination of contributions from the customer and the firm, which raises fairness questions in terms of whether the consumer or firm is granted ownership. Another challenge of data property ownership rights is that it lacks "a shared global understanding."[47] Moreover, nations struggle to balance data ownership rights "with government interests in the security of its citizens, national sovereignty and the need to use new types of data for legitimate state purposes including law enforcement."[48] In addition, "owning personal data might incentivize poor and more vulnerable people to sell their personal data, exacerbating existing inequities."[49]

Another pressing issue concerning data privacy involves international data sharing. Besides the sharing of data between individuals and an entity or between two different entities within the same territory, data can also be transferred across borders. As cross-border transfers are becoming more common, "the security of personal data transferred across national borders has been one of the drivers for international consensus on the fundamental principles for the protection of personal data."[50] "Prevalent factors cited when determining whether, and to what extent, to enact data localization laws include privacy concerns, international trust, and protectionism."[51] Nonetheless, there is still "uncertainty regarding data protection standards in foreign countries, [and thus,] many countries limit extraterritorial transfer of personal data."[52] Such limitations are known as data localization laws and regulations.

The emergence of data localization regulations "will likely generate complex conflicts between the requirements of the regulations from differing jurisdictions,"[53] and it will be a challenge, as well as likely costly, for companies to keep track of the different regulations and ensure compliance. Controlling international data sharing likely will be a particular challenge for companies participating in the Metaverse, as data sharing will be facilitated while the contributions of data from various locations may be difficult to fully account for.

Looking ahead, the attitude of different cultures toward data privacy is important, primarily because we are still unclear whether the Metaverse will be set up in such a way that borders will continue to exist. If borders will no longer be a concept, and one set of laws and regulations will cover the entire Metaverse, it could be a challenge to balance out the different attitudes toward privacy while trying to maximize overall welfare. Given how interoperable the Metaverse may be and how much easier the data-sharing process likely will become, standardizing data privacy rules at a global level based on the type of data being collected could prove to be the optimal, if not necessary, approach.

7

GOVERNING THE METAVERSE

This metaverse is going to be far more pervasive and powerful than anything else. If one central company gains control of this, they will become more powerful than any government and be a god on Earth.

~Tim Sweeney, Epic Games[1]

If anyone asked Mark Zuckerberg whether he aspired to be a "god of the Metaverse" in the way that Jon Postel was known as the "God" of the Internet,[2] we could not find a record, though he did state in 2021 that he wanted Metaverse users to be able to "build [their own] heaven."[3] Most likely, and especially after the Spring of 2023, AI would feature prominently in his response. Still, there is no getting around the fact that Zuckerberg renamed Facebook "Meta" in 2021 with a clear vision that Metaverse tech would feature prominently in the company's future and, more broadly, in the next iteration of cyberspace itself.[4] There can be tremendous economic benefits, after all, to being a first mover. Witness the dividends that the leading chipmaker Nvidia has been able to command given its dominant position in AI computing powered by Nvidia's industry-leading GPUs and associated software platforms.[5] Other leading tech firms, such as Apple, have become so dominant for so long thanks to their ownership of platforms,[6] which could soon include spatial computing.[7] These corporate giants have the potential to dramatically shape Metaverse governance, but so too do governments, which begs fundamental questions that were introduced in the preceding chapters, such as: What laws already govern the Metaverse? And how is, or should, it be governed differently from the Internet? What lessons can we apply from past experiences of regulating technology to ensure against tragedies of the Metaverse commons? This chapter addresses these topics and more.

What laws govern the Metaverse?

As explored in Chapters 3–6, myriad domestic and international laws, not to mention industry norms and codes of conduct, shape Metaverse governance. These include, in no particular order: intellectual property and copyright laws, contracts, torts, defamation, cryptocurrency regulation, tax, state-level privacy laws, sector-specific cybersecurity regulations, and Europe's General Data Protection Regulation (GDPR), to name a few.[8] These laws will shape some of the most likely Metaverse use cases, including in the areas of commerce, entertainment, and education.[9] For example, intellectual property cases involving the Metaverse have already been litigated. One such case involves the record label Roc-A-Fella suing Damon Dash, one of its co-founders, "seeking to enjoin him auctioning a NFT of the cover of the Jay-Z album *Reasonable Doubt*."[10] Such virtual assets will be regulated by a range of securities, property, banking, transmission, commodities, and tax laws.[11] As Metaverse games increase in popularity, there is also the likelihood that gambling and lottery laws will come into play; already some jurisdictions have probed the "loot boxes," which are "virtual unopened treasure chests," won by players.[12]

In general, any law or norm shaping Internet governance will also apply to the Metaverse given the extent to which these persistent, immersive, virtual communities (including Apple's spatial computing technology) rely on the Internet backbone.[13] The relative importance of these areas of law will doubtless wax and wane as Metaverse fads come and go, such as the digital asset and NFT craze of 2022, which drove interest in copyright laws.[14] But certain issues—such as privacy, identity, and cybersecurity—will persist and will only become more pressing as the Metaverse evolves, including whether predictions that spatial computing will inevitably lead to a "Metaverse 2.0" come to pass since the tech allows for a "more immersive, interactive, and intuitive digital environment."[15]

One, perhaps unexpected, application of Metaverse technology that could drive further government regulation is national security. A 2023 RAND report, for example, highlights the relevance of the Metaverse to advance the Department of Homeland Security's mission, such as with regard to fostering information sharing and improving training and preparedness, even as the technology also threatens to worsen concerns over misinformation, abuse, and cybersecurity.[16]

How does Internet governance apply to the Metaverse?

We take a lot for granted, including on the Internet. Many of us have never stopped to think about *how* the pulses of light on fiberoptic cables that carry your keystrokes and clicks around the world are regulated, and by *whom*. It may come as some surprise, for example, that a nonprofit corporation based in southern California manages the global domain name system, which matches cumbersome Internet Protocol addresses with easier-to-understand website names. Surveying all the decisions that got us to the current state of what passes for Internet governance is beyond the scope of this section, but a brief overview is warranted to provide some context.[17]

As the Cold War was ending in the early 1990s, there was a digital euphoria brewing about the potential of the Internet to become a borderless and self-governing space devoid of the authority of territorial states.[18] Famously, former US President Bill Clinton quipped that censoring the Internet was as tough as nailing "Jello to the wall."[19] Yet governments around the world, including perhaps most famously China, have successfully done just that. In China's case, early attempts at censoring the Internet were supercharged by the events of the 2011 Arab Spring (which underscored the potential of social networks to fuel unrest) by clamping down on anonymity, virality, and impunity.[20] Similarly, Russia's growing isolation in the aftermath of its 2022 invasion of Ukraine, and deepening ties with China, has seen its approach to Internet sovereignty evolve in a similar fashion. This has resulted in a growing "digital divide," not between those with and without broadband Internet access, but between those for whom the Internet is either "open" or "closed." The rise of cyber sovereignty, exemplified in efforts by India in 2018 and the United States in 2024 to ban TikTok and other Chinese-owned applications, showcase the extent to which cyberspace is no longer a global, open, secure, and interoperable network of networks (if it ever was).[21]

But what is "Internet governance?" One influential definition dates back to the 2005 World Summit on the Information Society, which defined the term as "the development and application by governments, the private sector, and civil society, in their respective roles, of shared principles, norms, rules, decision-making procedures, and programs that shape the evolution and use of the internet."[22] Although a helpful starting point, the term remains confusing given the extent to which technical standards setting is conflated with the emergent behaviors of Internet users. There is also no foreordained requirement that Internet governance must be multistakeholder in nature (i.e., involving representatives from governments, the private sector, and civil society). Indeed, it is the stated preference of a growing number of nations for multilateral (state-on-state) Internet governance. What is clear, though, is that an integral component of Internet governance is the development of norms for Internet operators and, increasingly, content providers along with technical standards underpinning internetworked systems.

In broad brushstrokes, the story of Internet governance may be broken down into three phases. Phase One encompassed influential network engineers and the ad hoc organizations that they developed, such as the Internet Engineering Task Force (IETF), extending from roughly 1969 to the birth of the Internet Corporation for Assigned Names and Numbers (ICANN) in 1998. Phase Two coincided with the commercial success of the Internet and the first global "digital divide" represented by the economic divergence of information and communication technology resources between developed and developing nations, which is the focus of this chapter. It also marked the emergence of so-called Web 2.0, which was first suggested in 1999 to describe the emergence of collaborative tools and online communities that generate content as opposed to the passive viewing of material online (i.e., Web 1.0). Finally, Phase Three has been defined to date by the extent to which nations have begun to assert a greater role in Internet governance, at least in their own countries, underscoring the potential for a "new 'digital divide'" to emerge not between the "haves and have-nots," but between "the open and the closed."[23] Yet, at the same time there are growing calls to focus on the emergence of a decentralized Web 3.0 (a term coined in 2006) built on distributed architectures that enable users to "control their own data, identity, and destiny."[24] Needless to say, such a techno-utopian vision is at odds with the increasing degree of cyber sovereignty being practiced by many nations, as was mentioned above.

There is a school of thought that Web 3.0 technologies, including those underpinning the Metaverse, could make existing stakeholders, such as ICANN, obsolete.[25] But such an outcome is unlikely, to say the least. True, ICANN's remit does not extend to blockchain-enabled Web 3.0 technologies, but given that domain names, user-friendly digital wallets, and digital identities remain integral to navigating the Metaverse, it's more likely that Web 2.0 and 3.0 decentralized governance structures will continue to coexist. This makes questions of governance that much more complex, including the role that companies are playing in shaping the Metaverse.

How are companies already governing the Metaverse?

When you purchase a digital asset, who owns it? You might think the answer is simple—the user who forked over the money, or cryptocurrency. But not so fast. Even though you may have purchased an NFT on a Metaverse platform that now resides in your digital wallet, which boasts a private key, when you dig into the terms of service, you will find that "all visual and functional aspects of digital assets" are "controlled by the private metaverse platforms and are subject to their unilateral control."[26] It should come as no surprise that often very few users bother to read the fine print, though. One study, for example, found that only 1.7% of users identified a "child assignment clause" that gave their first-born children to a fictional online service provider.[27]

This example makes clear that, as with cyberspace itself, companies are taking a central approach in determining the look, feel, and content of the Metaverse. This includes defining the terms of service around acceptable behavior, privacy, platform security, and dispute resolution.[28] But such choices involve difficult trade-offs, such as between interoperability, privacy, and security that are only beginning to be discussed and standards developed.

A variety of hot topics in corporate governance of the Metaverse are emerging. These include intellectual property rights management which has also become a hot topic, with companies actively monitoring Metaverse platforms for IP infringement, along with ensuring that virtual assets traded on the Metaverse meet the applicable securities, banking, and commodities laws.[29] Companies are also experimenting with "future proofing" their contracts by including clauses ensuring that third-party and existing relationships and assets could be deployed in spatial computing and Metaverse platforms[30] along with registering their trademarks in the Metaverse.[31] Finally, firms are exercising governance through product design choices, such as creating interoperable hardware and software based on uniform standards such as the World Wide Web (WWW) Consortium's XR Accessibility User Requirements (see Table 7.1).[32]

There is also some experimentation with applying governance models from related contexts to the Metaverse, such as Meta's Oversight Board that has taken a leading role in moderation (though not always without some controversy).[33] Some efforts have been made by regulators to oversee such initiatives (such as China as mentioned in Chapter 1) along with the European Union's Digital Services Act and Code for Disinformation, which could in time be extended to the Metaverse and are expanded upon in the next section. Indeed, the EU is actively investigating the form that Metaverse regulations should take, including with regard to the security of the underlying digital infrastructure and how that plays into existing regimes such as GDPR, as they project the global Metaverse market to reach $5 trillion by 2030.[34] Among other ideas is the potential to encourage the development of industry coalitions and codes of conduct similar to the applicable regime under GDPR.[35]

Table 7.1 Corporate Efforts to Regulate the Metaverse

Type	Description	Example/Parallels with Cybersecurity
Platform Rules & Policies	Companies establish guidelines for user behavior, content moderation, and data privacy on their platforms.	Meta Code of Conduct for Virtual Experiences
Self-Regulation	Companies adopt measures to ensure ethical conduct within the Metaverse, combatting issues like hate speech and misinformation.	Siemens Charter of Trust; Cybersecurity Tech Accord
Collaboration & Partnerships	Companies collaborate with governments, nonprofit organizations, and academic institutions to develop shared governance frameworks.	Christchurch Call to Eliminate Terrorist and Violent Extremist Content Online; Paris Call for Trust and Security in Cyberspace
Technological Solutions	Companies invest in tech solutions to improve governance, such as tools for detecting and removing inappropriate content or addressing privacy concerns.	AI Labeling; 2022 EU Code for Disinformation
Public Input & Feedback	Companies engage with users and the public to gather opinions and concerns about Metaverse governance.	NIST Cybersecurity Framework; NIST Privacy Framework

Yet Metaverse platforms also permit the potential for decentralized governance structures through so-called decentralized autonomous organizations (DAOs), which issue NFTs that give owners economic and voting rights on various issues.[36] Already, DAOs are being used to set the rules for the buying and selling of digital assets including cosmetics and real estate as NFTs, though in theory the same mechanism could be used for questions of content moderation and other vexing governance challenges.

These efforts are to an extent being guided by the efforts of civil society groups, such as the World Economic Forum, which are creating guidelines for firms to follow.[37]

Consider an experiment run by Metaverse and Stanford University, the results of which were announced in July 2023. In it, Meta undertook a deliberative, democratic process involving "over 6,000 people who were chosen to be demographically representative across 32 countries and 19 languages."[38] Participants spent hours in small-group conversations learning from non-Meta experts about the issues in play such as anti-bullying and moderation policies. The results looked promising, with some 82% of users suggesting that the company use this format for navigating some of the most challenging questions of Metaverse governance.[39] Meta was so encouraged, that it reportedly plans to roll out a similar process to guide its use and rollout of generative AI products and services.[40]

All this begs the question, though, as to whether companies, including Meta, should be able to govern themselves and their Metaverse platforms and properties. If Boeing is allowed to certify that a crash-prone aircraft is safe, and Facebook is permitted to violate users' privacy expectations, should companies and industries ever be allowed to police themselves?[41] The debate is heating up, particularly in the US tech sector, with growing calls to regulate the Metaverse and AI.

Are we asking for a tragedy of the Metaverse if we don't get this right? What lessons can we learn from other digital and real-world contexts?

It turns out to be possible, at least sometimes, for companies and industries to govern themselves, while still protecting the public interest. Groundbreaking work by Nobel Prize-winning political economist Elinor Ostrom and her husband, Vincent, found a solution to a classic economic quandary, in which people—and businesses—self-interestedly enrich themselves as quickly as possible with certain resources, including personal data, thinking little about the secondary costs they might be inflicting on others.[42]

To understand the idea behind their solution, first some context is in order. In a classic economic problem, called "the tragedy of the commons," a parcel of grassland is made available for a community to graze its livestock. Everyone tries to get the most benefit from it—and as a result, the land is overgrazed. What started as a resource for everyone becomes of little use to anyone, hence Garrett Hardin's famous insight that "freedom in the commons brings ruin to all."[43]

For many years, economists thought there were only two possible solutions to the scenario. One was for the government to step in and limit how many people could graze their animals. The other was to split the land up among private owners who had exclusive use of it and could sustainably manage it for their individual benefit. The Ostroms, however, found a third way. In some cases, they revealed, self-organization can work well, especially when the various people and groups involved can communicate effectively. They developed design principles to help understand the dynamics involved, along with furthering the concept of "polycentric governance."

For those new to the topic, the field of polycentric (multicentered) governance is a multilevel, multipurpose, multifunctional, and multi-sectoral model,[44] which has been championed by scholars, including Elinor and Vincent Ostrom. According to Professor Michael McGinnis, "[t]he basic idea [of polycentric governance] is that any group . . . facing some collective problem should be able to address that problem in whatever way they best see fit," which could include using existing governance structures or crafting new systems.[45] This robust model challenges orthodoxy by demonstrating both the benefits of self-organization, understood here as networking regulations "at multiple levels," and the extent to which national and private control can coexist with communal management.[46] It also posits that, due to the existence of free riders in a multipolar world, "a single governmental unit" is often incapable of managing global collective action problems. Their work can help determine if and when companies can effectively regulate themselves—or whether it's best for the government to step in, as many are doing already in the Metaverse context, as discussed in the next section.

All these factors can help predict whether individuals or groups will successfully self-regulate, whether the challenge they're facing is climate change, cybersecurity, or anything else. Trust is key, as Lin Ostrom said, and an excellent way to build trust is to let smaller groups make their own decisions. Polycentric governance's embrace of self-regulation involves relying on human ingenuity and collaboration skills to solve difficult problems—while focusing on practical measures to address specific challenges.

Polycentric regulation is a departure from the idea of "keep it simple, stupid"—rather, it is a call for engagement by numerous groups to grapple with the complexities of the real world. Yet self-regulation does have its limits—as has been clear in the revelations about how the Federal Aviation Administration allowed Boeing to certify the safety of its own software. Meta has also been heavily criticized for failing to block an anonymous horde of users across the globe from manipulating people's political views. In the face of public skepticism, firms like Meta and Apple have a stronger imperative to show that they can be trusted. Ostrom's ideas suggest they could begin to do this by engaging with peers and industry groups to set rules and ensure they are enforced, such as through the EU Code for Disinformation process.

So far, though, the track record is not encouraging. Failures of collective action are still all too common both online and offline. Examples range from unregulated areas featuring relatively undefined property rights, enforcement problems, and overuse issues, as with spam and Distributed Denial of Service (DDoS) attacks in cyberspace,[47] to "space junk" that is threatening the sustainability of low-Earth orbit. Indeed, one 2018 incident highlighted this danger with the International Space Station reportedly being damaged by a piece of orbital debris requiring the astronauts to use "duct tape to cover the hole . . . ."[48] Unfortunately, though, there are some problems that even duct tape cannot fix, given the larger problem of space junk threatening satellites and potentially hindering "space commerce, space tourism, [and even] the scientific exploration of space . . . ."[49]

Effective self-governance may seem impossible in the Metaverse because of the scale and variety of groups and industries involved, but polycentric governance does provide a useful lens through which to view these problems. Ostrom has asserted that this approach may be the most flexible and adaptable way to manage rapidly changing industries. It may also help avoid conflicting government regulations that risk stifling innovation in the name of protecting consumers without helping either cause. Yet Google has argued that while self-regulation is vital, it "is not enough."[50] Big questions remain, from regulating terms of service to tax policy. For example, should you pay sales and/or income tax on your digital assets? And to which jurisdiction? As such, governments can play an invaluable coordinating function and regulatory check on any excesses of Metaverse pioneers, which we explore next.

What role should governments have in regulating the Metaverse?

Governments around the world have no shortage of pressing concerns on which urgent action is needed. As of this writing, wildfires were raging out of control in Canada, burning an area larger than the state of Virginia. The United States saw an increase of 43% in the number of people killed by firearm violence between 2010 and 2020, though there was a thankful downturn in 2024.[51] Social media continues to be a negative and damaging influence on too many youth. Healthcare costs remain too high and access too limited. The scourge of drug addiction is scarring too many communities and families. Childhood poverty remains alarmingly prevalent. Inflation is persistently high. The list goes on.[52] So why should governments spend their limited legislative bandwidth regulating something as amorphous as the Metaverse?

Many countries are understandably answering the call with silence, but there are pockets of activity driven by key factors including the potential of Metaverse technology to help drive economic growth and technological leadership.[53] Admittedly, such promises can come off as illusory at best, and at worst an excuse for overly lax regulation. But the prospects for Metaverse applications (and their abuse) in contexts such as housing construction and real estate walk-throughs demand regulatory attention, in addition to training healthcare professionals in virtual or augmented environments that would include their patient's biometric information that could be updated in real time. Governments also likely will have some role to play in ensuring that—similar to broadband Internet access—as Metaverse technologies such as spatial computing (or, as Google refers to it, "ambient computing") mature,[54] they are widely accessible as part of a broader push towards twenty-first century digital literacy and digital citizenship. Exploring the potentials and perils of this technology in educational settings will be important to this end, including by offering training programs for AR/VR content creators, future designers, engineers, and architects. Lastly, governments are already experimenting with using the Metaverse to open digital embassies and promote cross-cultural dialogue, as discussed in the introduction of this chapter. Such cooperation would hold the potential to promote the betterment of international relations. Imagine a school child from Indiana walking alongside a student of the same age from Tuvalu and developing firsthand empathy for each other's climates and cultures as well as the challenges they face.

In other words, governments have the potential to play many important roles in fostering the rollout of equitable, inclusive, useful, and innovative Metaverse platforms that safeguard free speech and privacy rights along with promoting cybersecurity. Of particular concern in the US context have been issues of content moderation, data privacy, market power and competition, and addressing the digital divide, given that more than 14 million Americans still did not have access to broadband Internet as of 2019.[55] The Congressional Caucus on Virtual, Augmented, and Mixed Reality (MR) Technologies (known, perhaps, somewhat ironically as the "Reality Caucus") has been focusing on these issues for years; indeed, among the first proposed bills that would have regulated VR for medical procedures dates to 1992.[56] Augmented, mixed (so-called XR, which could include Apple's spatial computing efforts), and virtual reality applications, though, are just some of the underlying technologies shaping the Metaverse, which also relies on 5G/6G deployment, and blockchain/NFTs.[57] Indeed, as the Congressional Research Service notes, there are differences of opinion, discussed throughout this book, about whether the Metaverse represents a fundamental change in the architecture and operation of the Internet itself or merely an evolution with targeted appeal—the notion of so-called "proto-metaverses" such as massively multiplayer online (MMO) games and virtual concerts.[58] The bursting of the Metaverse hype bubble in early 2023 will likely decrease perceived pressure on regulators as their focus shifts to AI, but the underlying issues related to governing immersive, persistent, interoperable networks across platforms are not going away. Of particular interest is the phenomenon known as the "embodied Internet" made possible by users' positive experience of being present in an online environment rather than being a passive observer, underscoring issues of identity and, relatedly, mental health as explored in Chapter 5.

The immersive, expansive, persistent, and potentially global nature of the Metaverse means that content moderation is especially challenging (consider the Section 230 debates about platform liability for third-party content in this context), particularly in the areas of user identity protection and managing misinformation. The immersive nature of the Metaverse makes some issues like bullying and harassment that much more concerning,[59] while also ballooning the amount of data provided by users in these environments (including, potentially, neural information) that is already subject to, at best, fragmentary safeguards.[60] Yet in contrast to Europe (discussed next), it is unlikely we'll see a comprehensive piece of US legislation to manage the myriad regulatory issues facing Metaverse providers and users, as is the case for an omnibus federal law defining privacy rights or cybersecurity requirements. Instead, targeted reforms dealing with specific issues are much more likely. For example, the proposed Immersive Technology for the American Workforce Act would create a five-year grant program administered by the Department of Labor to support community colleges along with career and technical centers "in developing education and training programs for workforce development utilizing immersive technology."[61] Other areas that could lead to bipartisan movement include the need for upgraded infrastructure to improve the capacity of existing broadband networks for widespread use of spatial computing and other immersive technologies, especially given prevalent problems of latency in DSL and cable networks.[62] International, multi-stakeholder engagement is also key to developing "open standards and globally accepted protocols," as was so instrumental in the update of TCP/Internet Protocol.[63]

It should come as little surprise, given its leading role in data governance, that among the most proactive jurisdictions around the world in regulating the Metaverse is the European Union. In particular, the EU has created the 2030 Digital Compass that "sets out to improve the digital skills of the European citizens and emphasize the development of more skilled digital professionals; transform the EU's digital infrastructure landscape; and enhance European cybersecurity by improving data infrastructure to enable data to remain within the EU and not be stored in third countries."[64] As a result, rules for data localization look set to only be further entrenched in the future, calling into question whether a true, global Metaverse (similar to considering cyberspace as a "global networked commons") is possible or desirable. Indeed, this may be considered a component of Europe's drive for "digital sovereignty" by 2030. This is a goal mirrored by the efforts of other nations around Internet and cyber sovereignty, which may be read as a challenge to the US-led vision of cyberspace—and, relatedly, the Metaverse—as a global system that is "open, free, global, interoperable, reliable, and secure."[65]

Government involvement can help build bridges, break down silos, and solidify trust across Metaverse stakeholders, as happened with cybersecurity efforts from the National Institute for Standards and Technology. Polycentric governance can be a flexible means to this end, adapting to new technologies more appropriately—and often more quickly—than sole governmental regulation. It also can be more efficient and cost-effective, though it's not a cure for all regulatory ills. And it's important to note that regulation can spur innovation as well as protect consumers, especially when the rules are simple and outcome focused.

Consider the North American Electric Reliability Council (NERC). That organization was originally created as a group of companies that came together voluntarily to guard against blackouts. NERC standards, however, were eventually made legally enforceable in the aftermath of the Northeast blackout of 2003. They are an example of an organic code of conduct that was voluntarily adopted and subsequently reinforced by government. To the extent that such a sequential approach is optimal, it would ideally not require such a crisis to spur this regulatory process forward in the Metaverse context.

Ultimately, what's likely needed—and what Professor Ostrom and her colleagues and successors have called for—is more experimentation and less theorizing, especially as new technologies (such as AI and brain-computer interfaces) promise an even more immersive Metaverse experience than AR/MR/VR, which is explored further in Chapter 8.[66]

8
AI IN THE METAVERSE

The Metaverse can be summed up as the Internet in three dimensions.

~Deloitte Chief Disruptor Ed Greig[1]

The significance of artificial intelligence (AI) on nearly every aspect of modern technology cannot be understated. Tech analysts and economists have even argued that AI and big data are the foundations of humanity's fourth industrial revolution.[2] Recent advancements in generative AI have further cemented the role of artificial intelligence as a civilization-scale, transformational technology. Many creative roles previously considered safe from automation are now in danger of being usurped by significantly faster and cheaper AI programs. Authors, artists, musicians, and videographers—as well as the companies that employ them—are all feeling the economic pressures presented by innovations in AI. The impacts of AI on the Metaverse are similarly expansive, as explored in this chapter.

How are AI and the Metaverse related?

Because of the pressing economic and cultural impact that AI has recently had, some tech pundits have proclaimed: "The Metaverse is Dead, Long Live Generative AI!"[3] Investors, too, have changed their priorities.

Between 2022 and 2023, first-quarter venture capital funding for Metaverse companies fell from $2 billion to $587 million. At the same time, first-quarter venture capital funding for generative AI companies grew from $613 million to $2.3 billion.[4]

Despite these proclamations, the idea that the Metaverse has been left in the dust as yesterday's tech fad misses the critical link between the Metaverse and AI. As with all other Metaverse technologies, AI is a foundational pillar of any vast, interactive, and ever-growing implementation of the Metaverse. By integrating AI and the Metaverse, the utility of both technologies in the realms of art, education, communication, entertainment, medicine, and commerce is exponentially increased. Generative AI has a tremendous ability to deliver virtual worlds at scale by creating realistic and diverse virtual characters, environments, objects, text, and even audio.[5]

AI accelerates—and makes financially viable—one of the most challenging aspects of building the Metaverse: making dynamic and adaptable content.[6] AI does so "by providing some of the most tedious, butnecessary, building blocks of virtual worlds."[7] At the same time, the combination of AI and the Metaverse risks creating the "next generation of privacy-violating, competition-thwarting, and truth-killing platforms," as Tom Wheeler wrote for The Brookings Institution.[8]

Beyond the synergy between these technologies (and the possible dangers derived from it), there is also a conceptual link between AI and the Metaverse. As AI permeates the fabric of society, we find ourselves struggling to define what exactly is artificial about artificial intelligence. As the Metaverse continues to grow, we face a very similar question: What does the virtual in virtual reality really mean? Is there a truth when engaging with AI? Is there reality when engaging with the Metaverse? Answering these questions becomes even harder when AI and the Metaverse converge. Although this chapter will leave these larger philosophical questions unanswered, the relationship between AI and the Metaverse is a deep one, and this chapter begins to dissect how the two technologies will interact.

Who and what is real in the Metaverse?

Before diving into the complex questions of what reality means when AI meets the Metaverse, it behooves us to first discuss what reality means in the Metaverse by itself. Unfortunately, despite being the "R" in the names VR, AR, MR, XR, etc., the concept of "reality" is somewhat difficult to define.

In popular parlance, the Metaverse is frequently discussed in contrast to reality, much like the Internet is discussed in contrast to "real life." The abbreviation "IRL," which stands for "in real life," was even popularized to distinguish online interactions from reality. But framing the discussion about the internet as mutually exclusive from real life mistakenly suggests that the internet is unreal, false, or fake. This conceptualization suggests that the environments we inhabit online are fictional, and that the personas we exhibit online are fanciful if not fraudulent.

Social media, however, has already shown us that this perspective on reality is flawed. As the role of social media in everyday life continues to expand, our online activities and digital presence have become an integral part of our reality.[9] Consequently, the Internet has become a "psychological space; an extension of our individual and collective minds."[10] Even when our digital presence is fragmented across various platforms, websites, and devices, "it seems clear that our various online personas are all digital breadcrumbs of the same persona; different symptoms of our same core self."[11]

The Metaverse, like social media platforms, is not a fictional digital environment in which our true selves are left behind. Rather, the Metaverse is a true extension of our parallel realm in which real interactions, emotions, and experiences unfold. In the Metaverse, real people work, socialize, and even create. Real economic transactions take place. Real crimes occur. Real relationships are formed. Ultimately, the Metaverse is not a departure from reality but an extension of it. Beyond these social and economic realities, advancements in technology, too, help dissolve the idea that reality refers solely to physical space. For example, by tracking "[e]ye movements, heart rate, facial expressions, and even perspiration," XR technologies make our presence and identity in the Metaverse that much more indistinguishable from and entangled with our physical reality.[12]

As discussed in Chapter 5, the Metaverse is not like a video game in which players dive into a bounded and explicitly fictional world. The Metaverse is not about discarding one's identity and adopting a new character's personality, looks, and behaviors. Rather, the Metaverse, much like social media, encourages users to embody their profiles or avatars, immerse themselves fully into each virtual environment, and interact seamlessly with their surroundings. The environments, personas, and actions taken online are as real and as consequential as anything else we do "IRL." Given these factors, the Metaverse cannot be categorically labeled unreal or fictitious.

Nonetheless, not everything on the Internet and the Metaverse is real. Certain things are uncontrovertibly "fake." Fraudulent activity, for example, whether online or in real life, is a clear example of fake content (almost by definition). In the context of social media, fraudulent behavior takes on various forms. The term "catfishing," for example, refers to a particular kind of fraudulent activity in which an individual (called the "catfish") creates a fake online profile on social media platforms, frequently employing stolen pictures, backgrounds, and personalities. If successful, the catfish deceives victims into unwittingly revealing secrets, sharing resources, and buying into further falsities. This "deceptive practice" of catfishing is used "for anything ranging from pranking ... friends to carrying out serious crimes, like sexual assault and identity theft."[13] Just because our interactions online are real and consequential, does not mean that everything and everyone online is truthful and that everything is as it seems. Such falsities unfortunately abound.

Phishing is another example of clearly fraudulent online activity, as discussed in Chapter 4. As the Federal Trade Commission summarizes, "[p]hishing emails and text messages often tell a story to trick you into clicking on a link or opening an attachment. You might get an unexpected email or text message that looks like it's from a company you know or trust, like a bank or a credit card or utility company. Or maybe it's from an online payment website or app."[14] Victims of phishing are tricked into entering login credentials, disclosing payment information, and even installing malware through these fraudulent messages. Much like catfishing, phishing is also a deceptive practice, perpetuating falsities in our online environments.

What do these two examples have in common? What makes them representative of unreal behavior? It is not that these activities take place online. Rather, these examples involve purposeful deceit. It just so happens that such deceit is enabled or facilitated by the online environments that have become such foundational aspects of our everyday lives.

Much like social media, which has enabled these novel methods of fraudulent behavior, the Metaverse too will facilitate new ways of enacting deceit. By relying on representative avatars, 3D rendered environments, and digitally mediated communication, the Metaverse may exacerbate our vulnerabilities to these types of falsities. Our deeper perceptual immersion in the Metaverse may, for example, lower our defenses against suspicious-looking emails or unexpected friend requests.

Delineating reality in the Metaverse becomes even more difficult as Metaverse creators, users, and abusers come to wield the power of AI. Deep fakes, chatbots, and other artificial intelligence applications make it far easier to deceive innocent users both on the wider social internet and especially in the Metaverse. Such technologies, for example, may make it virtually impossible to distinguish between human-avatars and those controlled by AI. These AI-controlled bots may have legitimate uses in the Metaverse, as discussed later in this chapter, but they may be far too easy to abuse. Metaverse bots that can "emulate physical and emotional characteristics like facial expressions and body language"[15] may, for example, trigger far stronger subconscious predispositions of trust.

In social media, deepfakes, or "fictional AI-generated photos and videos, have already begun to flood the internet, sparking controversy and confusion among everyday users and mainstream media."[16] In the Metaverse, deepfakes will only become more realistic as generative AI facilitates the creation of picture-perfect avatars with human-like speech and motions.[17] Powered by artificial intelligence's ability to collect and analyze users' physiology, the Metaverse might even allow catfishers to create virtual identities that perfectly "emulate [the] physical and emotional characteristics" of real individuals.[18] Additionally, artificial intelligence in the Metaverse can monitor users' interactive behaviors, generating valuable and sensitive data for phishing attacks.[19]

Determining what is real in an AI-driven Metaverse may be more difficult than in any other digital environment we have faced so far.

Does AI undermine reality in the Metaverse?

As previously discussed, the Metaverse is part of reality. As such, it hosts real friends, real art, real commerce, and unfortunately, real deceit. Does the prevalence of AI in the Metaverse tilt the scale from reality to fiction? In other words, does AI undermine the reality of the Metaverse? To answer this question, let us first determine the various legitimate uses of AI in generating the Metaverse. Specifically, let us look at how avatars and environments will all be shaped by the rise of AI in the Metaverse.

How does AI shape avatars?

To exist in and interact with the Metaverse, individuals rely on an avatar, a 3D rendered depiction of one's body. It is through this avatar that we emote, communicate, and behave in the Metaverse. It is also through this avatar that others perceive us. In the physical world, our facial expression, body language, and even level of perspiration may also signal to others our level of stress, engagement, or empathy. Unfortunately, in the early implementations of the Metaverse, avatars were far too simplistic to be effective means of conveying the full dynamic range of one's true self in the Metaverse.

Take the first two generations of Mark Zuckerberg's very own Metaverse avatar shown in Figure 8.1.[20]

On the left is Meta's first generation so-called "dead-eyed avatar."[21] These images of avatars received a significant amount of backlash due to

Figure 8.1 Evolution of Zuckerberg's Avatar

their lack of realism, facial expressions, and even legs. Op-eds were written about the "hideously ugly aesthetic" of this version of the Metaverse, which was perceived to be "completely devoid of texture and, more importantly, personality."[22]

On the right is Meta's second generation "upgraded" avatar, which featured a significantly more detailed depiction of Mark Zuckerberg. Despite these improvements, the new generation was also mocked. Columns were written noting the avatar's similarity to "the pale, vacuous look of a Victorian ghost, taking a selfie in front of basic renderings of the Eiffel Tower and the Basílica de la Sagrada Família."[23]

These derisive comments made against both generations of avatars are reproduced here for more than humor. Beneath these aesthetic judgments is a real critique about what Metaverse avatars must be able to convey to successfully achieve their purpose. Both generations of avatars lacked the realism necessary for true immersion in the Metaverse. Face-to-face virtual meetings, which is one of the primary use cases for the Metaverse, felt stunted with these avatars. By being unable to match the expressive level of our physical human communication, these avatars gave virtual connections a muted or even fictional tinge.

Figure 8.2 Zuckerberg's New Avatar

Built into these comments is the idea that "photorealistic avatars in virtual or augmented reality" are not only the gold standard for communications in the Metaverse, but they are also almost required.[24] With recent advancements in AI, it seems that achieving such realistic avatars is well within reach.

With the use of state-of-the-art AI, Meta has upgraded its avatars yet again. The latest virtual iteration of Mark Zuckerberg is shown in Figure 8.2.[25]

To build these new avatars, Meta originally relied on a system of 171 high-resolution cameras which fed data into four high-end GPUs which then rendered a virtual avatar after six hours of processing.[26] Clearly, this method—though successful—was far from scalable. Since then, through the development of novel AI methods, specifically new neural network machine learning (ML) algorithms, Meta has been able to generate avatars with nothing more than an iPhone.[27]

Similarly, Apple has built two systems of generative AI for avatar-generation. The first system of so-called digital "personas" is scheduled to be launched with the Apple Vision Pro. By using the device to capture one's face while raising eyebrows, smiling, and blinking, Apple's AI algorithms can stitch together a realistic 3D rendering of the user's real-time facial expressions.[28] The second system, which is designed for full-body analysis, can turn a short video or between 50 and 100 frames into a fully animated 3D avatar.[29] Because so little video is required, the AI generates aspects of the avatar "to fill in the gaps of what was captured and included in the model."[30]

As this series of iteratively more realistic avatars demonstrates, "one of the challenges we face in virtual and distributed collaboration is the absence of nonverbal cues and emotional nuances that form an integral part of face-to-face interactions. Generative AI has the potential to bridge this gap by imbuing virtual avatars with realistic emotions and body language, creating a more immersive and emotionally resonant experience."[31]

Thus, despite their existing successes, Meta, Apple, and countless other companies continue to invest in AI innovations to build photorealistic Metaverse avatars as fast as possible, from as little data as possible. If the Metaverse becomes home to millions or billions of users, many if not most of them will not have access to a "4-D photogrammetry booth (a huge frame with about a hundred digital cameras) to capture all possible facial details."[32] Even if they do, individuals will likely not want to sit for a 30-minute motion capture session for perfect precision in their avatar. Thankfully, it seems that AI is up to the task.

Note, however, that the use of AI for avatar-generation is not without controversy. The 3D biometric data required for AI-facilitated avatar generation is highly sensitive personal information that can be used to identify someone. As of this writing, such data is protected by what some may describe as a confusing patchwork of ineffective state-by-state regulations in the United States.[33] Discussions about privacy in the Metaverse found in Chapter 6 will play a significant role in defining the scope of AI-generated photorealistic avatars.

How does AI render environments and assets?

To appreciate how AI will drastically change the way in which Metaverse assets are created, it is first necessary to understand how such things (3D models, textures, icons, sound effects, etc.) were traditionally built for the video game industry.

Existing expansive virtual-world games—those that most closely mirror the Metaverse—can be incredibly expensive to build. Development costs between $80 and $300 million are now commonplace. A Bain & Co. analysis predicts that these costs may be heading toward $1 billion per game by 2027.[34] What exactly is this exorbitant amount of money being spent on? Main Leaf, a game development company, answered this question by noting that asset creation is "[o]ften the bulk of a game's price tag."[35] This development is also the step that takes the longest, the step in which game developers have "hundreds, if not thousands, of workers" working for "around 4 years."[36] Even once 2D art has already been created, "converting 2D sketches into 3D assets takes thousands of hours across a standard-sized video game."[37] To this day, asset creation remains a "manual and time-consuming" process and, unfortunately, "there is a limited supply of talented 3D artists."[38]

It is no wonder that game developers are extremely interested in using AI to decrease this workload (and its related costs). Creating concept art, 3D objects, and even entire landscapes are now subject to AI automation. So, many companies, all claiming to save time and money for game developers, are exploring use cases for AI in game generation that an entire development workflow can be built on generative AI technologies. Take the Seattle-based startup Rec Room as an example. As a research project, the team "visualized ideas via Midjourney and DALL-E, and turned the resulting images into 3D assets with CSM and Shap-E. Finally, . . . alien skies were built out with Skybox, an AI tool from Blockade Labs."[39]

Because the Metaverse also relies on 3D-rendered environments and assets, much of this innovation can be equally revolutionary for the Metaverse, if not more so. Beyond generating assets, AI tools can be trained to create (or recreate) photorealistic content for the Metaverse. As the Boston Consulting Group noted, "We are getting closer to scenarios where GenAI could be trained on existing photos, maps, and building specs and could create an immersive world that has the real and physical properties of the actual environment."[40] Jensen Huang, the CEO of Nvidia, agrees. "We can now do generative AI for images. We can do it for videos. At the rate that it's moving, you'll do it for entire villages; 3D villages and landscapes and cities and so on. You'll be able to assemble an example of an image and generate an entire 3D world. That's going to enable the metaverse like you can't believe."[41] AI technologies will also democratize content generation for the Metaverse. No longer will specialists with years of training be necessary to create such digital content. Instead, regular users will be able to rely on prompt-based generative AI technologies to build their own environments, assets, animations, and even avatars in the Metaverse.

Is AI-generated content real?

AI technologies are revolutionary in enabling a bigger, more immersive, and more interactive Metaverse. Is this bigger, more immersive, and more interactive Metaverse any less real because it was generated by AI? Regardless of the process, this AI-generated Metaverse gives real people, engaging in real activities more, not less, room to express themselves and live out their lives.

How will AI bots populate the Metaverse?

The Metaverse is a place built by humans for humans. Nonetheless, AI bots will likely have a large presence in the Metaverse. In social media platforms, which were also built by humans for humans, AI bots have already taken a stronghold.[42] The presence of AI on social media, however, is not inherently problematic. As will be discussed below, AI bots have their legitimate and useful purposes. Nonetheless, AI bots can also have disastrous consequences, many of which will be exacerbated in an immersive Metaverse. But to discuss how AI bots will affect the Metaverse, let us first understand how these discussions have played out in the world of social media. Specifically, let us look at automated social media accounts.[43] These accounts can be used for both legitimate and fraudulent purposes. AI bots in the Metaverse will present similar opportunities and challenges. Such challenges may be multiplied when AI bots become indistinguishable from human avatars.

A basic method of automating social media accounts was to create simple algorithmic bots that either reposted content generated by humans or relied on simple rules to adjust, paraphrase, or question such content. The rules and algorithms involved are also considered a version of AI despite lacking the more common forms of AI seen today, such as machine learning. This style of automated account still exists. In fact, the creation and maintenance of these bots is sometimes sold as a legitimate service. One company, for example, advertises that its bots can automatically send direct messages, tweet, retweet, and the like.[44] Unfortunately, these exact same technologies can be used to generate spambots, fake followers, and amplification bots, which "take advantage of Twitter [X] to spread fake news and false information."[45] The scalability of these simple bots presents a problem. If, for example, thousands of bots are "talking on a particular topic, or using a particular hashtag, it makes that thing look more popular than it actually is."[46] Thus, these simple automated bots can be effective tools for fraud and disinformation.

Newer automated accounts relying on generative AI technology, such as ChatGPT, can handle a wide variety of tasks that the earlier generation simply could not. Consequently, companies have promoted generative AI as "your next best ally for leveling up your [social media] game without breaking a sweat. This powerful tool can handle automated conversations on your behalf, freeing up your time for more important tasks."[47] The promoted legitimate use cases of generative AI include the following: providing customer service responses, generating sales leads, engaging in autonomous market research, and original content generation.[48] Unfortunately, as before, these exact same tools can be used to cause harm to individuals and communities alike. Without generative AI, disinformation campaigns were frequently easy to target and take down due to the repetitive nature of the posts. With generative AI, "the creation and spread of compelling fake news stories, social media posts, and other types of disinformation have become more accessible and cost-effective than ever before" and, unfortunately, more difficult to identify and prevent.[49]

Because of the harms posed by automated social media accounts, the US Cybersecurity and Infrastructure Security Agency (CISA) has subcategorized malicious automated bots into five groups in an attempt to inform the public about these dangers:[50] *like-farming bots* inflate engagement numbers, fabricating false popularity and traction; *hashtag-highjacking bots* spam users who are discussing certain topics or using certain hashtags; *repost storm bots* flood social media networks with malicious content; *watering hole attack bots* target top trending topics to attract unsuspecting audiences; and *sleeper bots* remain largely inactive and, therefore, undetected until they are used as one of the other categories of bot accounts.[51]

Combined, these malicious bots can be used for any number of nefarious purposes, such as:

- Interfering with democratic elections
- Manipulating financial markets
- Drowning out free speech
- Broadcasting spam and phishing attacks
- Fraudulently amplifying popularity

Nonetheless, it is important to remember that automated bot accounts are not categorically malicious. As the CISA acknowledged, bots can be used for "various *useful*" tasks.[52] As such, an outright ban on automated accounts would be detrimental to business innovation, customer service, and even the free exchange of ideas.

Imagine, now, AI-driven bots in the Metaverse. These are autonomously controlled avatars who wield the power of artificial intelligence to "engage and interact with people in the virtual world."[53] Such bots could also provide legitimate services in the Metaverse, such as customer service, automated information retrieval, data security, etc. For the purposes of immersion and consumer confidence, these bots would be made to act as naturally as possible. In other words, even for legitimate purposes, AI bots would be made to act as human as possible.

Nonetheless, these bots are obviously not human. They engage in "interactions with other intelligent entities, based on [their] understanding of both the goals of [their] creators and the capability or functionality of the simulation and the data enabling [them]."[54] This means that depending on the "goals" set for them, the very same AI bots could be trained to cause havoc in the Metaverse. AI bots may sow disinformation, cause public disturbances, and engage in fraudulent activities in the Metaverse.

Malicious bots in the Metaverse can take on another equally devastating form: deepfake avatars. These photorealistic avatars with perfect imitations of voices and movements may be used for the creation and spread of harmful content, such as "the manipulation of pornographic material (for example, revenge porn) and for political purposes (for example, to fake political statements or actions)."[55] The inability to distinguish between human avatars and AI bots in the Metaverse would facilitate these fraudulent activities and make the Metaverse a less safe, less trustworthy environment.

If—prior to generative AI—regulators thought they would be able to identify and disable problematic Metaverse bots because of their repetitive or robotic movement and speech, recent developments in AI have made this goal significantly more challenging. Yet, banning AI bots in the Metaverse is also problematic, if not impossible. AI bots will surely inhabit the Metaverse; the question to be answered is: what, if anything, can regulators and companies do to ensure sufficiently stringent guardrails, so that their presence does not undermine the entire purpose of the Metaverse?

What rights should AI bots have in the Metaverse?

Does AI have any legal rights? Does AI have legal personhood? These types of questions have been rich fodder for science fiction and pop science for generations.[56] Given the importance and widespread use of AI, it is not surprising that these questions have outgrown their origins. Legislators, courts, and legal researchers have all turned their attention toward defining the rights and duties of AI. Much of this discussion has taken place in the context of intellectual property. Who owns the inventions, texts, and art generated by AI? Is it the AI itself? The user of the AI? The creator of the AI? Or no one at all? Although these questions are of critical importance and have been raised in various jurisdictions across the world,[57] they are not unique to, or affected by, the Metaverse. Therefore, this chapter will not focus on the question of AI intellectual property rights. Rather, this discussion will highlight the question of free speech rights for AI bots precisely because the Metaverse raises a unique set of challenges and opportunities in defining such rights.

Unlike physical spaces, in which laws and regulations can now easily distinguish between human individuals and nonhuman entities, the Metaverse removes the distinction of embodiment between humans and AI. In this digital realm, both humans and AI interact with their environment through avatars, rendering humanity's physical presence irrelevant. As discussed above, it is even possible that generative AI may ultimately make algorithmically controlled Metaverse bots that are indistinguishable from human-controlled avatars. This technological shift raises critical questions about the role and rights of AI in the Metaverse, particularly concerning free speech. On the one hand, if humans and AI are indistinguishable, it may be impossible to reserve speech rights solely for humans. On the other hand, if AI-speech serves useful policy goals, it may be worthwhile to protect AI speech as AI speech per se. As will be discussed, however, once AI is granted speech rights, there could be unintended consequences in terms of what speech pervades the Metaverse.

Note that the discussion here does not refer to consciousness in artificial intelligence or the singularity. This analysis focuses on existing AI technologies as they may be implemented in the Metaverse. Note also that the discussion here bypasses the ongoing debate about whether free speech rights apply to privately owned social media platforms.[58] Currently, First Amendment-based free speech protections in the United States categorically do not apply to these "private public squares." Rather, the discussion here targets the policies and rationales behind granting any sort of speech protections to AI-generated content in the Metaverse. These protections can take the form of legal mandates, private policies, or anything in between.

There are various compelling reasons to consider extending free speech rights to AI in the Metaverse, but here is just one example. Granting speech rights to AI bots in the Metaverse may be useful in protecting censored, underrepresented, and minoritized speech. Individuals in the Metaverse are not immune from government or social sanctions. This is especially the case as avatars continue to grow more interlinked with our physical and legal identities. Therefore, the very same political, social, ethnic, and racial dynamics that exist in the physical world will extend to the digital one. Individuals may be pressured against sharing their legitimate perspectives. This chilling effect against speech may not only hinder individual autonomy, but it may also undermine the "marketplace of ideas" and even have negative implications for "democracy and self-governance."[59] Take the chatbots currently being developed in China as an example. Although these bots are required to comply with the country's content-based censorship laws, they nonetheless run the risk of running afoul of these regulations. What might these trained chatbots say "if prompted to discuss democracy, China's Constitution, Xi Jinping from the perspective of dissidents and [human] rights lawyers like Liu Xiaobo and Gao Zhisheng or Xi's intra-CCP rivals like Bo Xilai?"[60] Now imagine if a chatbot were programmed to purposefully make such political information available. AI bots could be a tool for political expression.

By employing AI bots in the Metaverse to relay information and opinions, including political speech, communities may have a chance to bypass current power dynamics that keep individuals from speaking their minds. In the same vein, first-time and underfunded political candidates could also rely on AI tools in the Metaverse to "lower financial barriers to entry" into the political process.[61] The impact of such outsourced speech may be higher in the Metaverse than almost anywhere else. If human-avatars and AI bots become indistinguishable, the Metaverse will be an environment in which the speech of all entities is granted an equal footing and receives equal attention.

It is because of this even playing field that extending free speech rights to AI in the Metaverse also poses serious dangers. Speech-bots may be used to exacerbate and automate harassment, multiply and spread misinformation, and undermine the ability of the Metaverse to create a useful and pleasant digital environment. In the context of political speech, AI can already be used to wreak havoc:

> AI-fueled programs, like ChatGPT, can fabricate letters to elected officials, public comments, and other written endorsements of specific bills or positions that are often difficult to distinguish from those written by actual constituents. These fabrications—and the speed and volume at which they can be created—may be used to generate the appearance of public consensus on a given issue and pressure legislators to act on a desired agenda.[62]

Now imagine all this taking place in the Metaverse with immersive and intimate, interpersonal, allegedly "live" communication replacing the letters, phone calls, and written comments created by current AI technologies. It may become impossible to gauge where a constituency truly stands on any political question. Voters may be misled about the existence of a community consensus. Photorealistic 3D deepfakes of political candidates may sow further disinformation and distrust into an already fraught electoral process. As previously noted, attempting to regulate AI speech may become impossible if identifying AI speech becomes impossible. But until then, there may be cause for significant constraints on AI speech rights in the Metaverse.

Ultimately, as we venture deeper into the Metaverse, the question of AI rights, particularly concerning free speech, poses a significant and complex challenge. While there are persuasive arguments for extending such rights to AI, particularly in fostering a rich and diverse discourse, the potential risks cannot be overlooked. This requires a careful and nuanced approach that balances the need for open and free communication with the imperative to defend against the spread of misinformation and democratic harm.

9
OUR META FUTURE?

> It's not safe out here. It's wondrous, with treasures to satiate desires both subtle and gross. But it's not for the timid.
>
> ~Q, Star Trek[1]

This book has explored the economic, privacy, security, political, and governance dimensions of the Metaverse. But, in many ways, the Metaverse is just one aspect of a hyper-connected digital future in which augmented humans will interact with blended real and virtual worlds in both work and play. How will this future look and feel? And what can we do to make it a good one in which we not only avoid the dystopian scenarios spun by science fiction authors but set the stage for new possibilities for human progress. This chapter explores lessons gleaned throughout the book to develop such a lens to help us look ahead, projecting technological and regulatory trends forward to visualize whether we're hurtling toward the Matrix or the holodeck, with stops along the way to consider how we can work together to craft a common—and humane—future both online and offline.

Inspirations and warnings from sci-fi: Are we hurtling toward Ready Player One, The Matrix, or something better/worse?

Pop culture's visions of the Metaverse range from dystopian nightmares (*The Matrix*) to escapist utopias (*Ready Player One*) to, well, holodeck poker in Starfleet uniforms. So, which of these futures are we hurtling toward? The key, as always, lies in our choices. Science fiction has an impressive history of accurately forecasting future technologies. Dick Tracy's wristwatch communicator? Check. Starfleet's flip communicators? Ditto. 1960s visions of online shopping and education? Nailed it. Chatbots, like ChatGPT, eerily echo *2001: A Space Odyssey*'s HAL's chilling sentience (minus the homicidal tendencies, hopefully).

Why are science fiction writers so good at predicting gadgetry? The answer is less magical than it seems. Sci-fi, you see, isn't just a telescope that helps us see the future; it's a beacon illuminating the path ahead. Planting the seeds of neat gizmos in impressionable young minds shapes the engineers and innovators who bring those possibilities to life. Witness Qualcomm, whose founders confess their debt to Starfleet's tricorder dreams. Or Igor Sikorsky, the inventor of the helicopter who credited Jules Verne's book, *Clipper of the Clouds*, as his inspiration.

Given this track record, science fiction likely offers realistic glimpses of the machines we'll be using in the future. Virtual environments that look and sound just like reality are likely in our future, and they will be delivered via attractive and unobtrusive wearable devices.

On the other hand, science fiction's ability to predict the societal ramifications of technology is a mixed bag. Take Jules Verne, who conjured the idea of submarines with a sense of wonder, yet it was devoid of foresight into their military ramifications. Similarly, while the advent of the Internet was foretold, the proliferation of social media and its attendant difficulties—ranging from misinformation to the erosion of public discourse—was in nobody's crystal ball. *Star Trek* gifted us with the vision of personal communicators, a prelude to our smartphones, yet it failed to see the gig economy and how it would reshape labor markets.

Why does science fiction frequently stumble when forecasting our relationship with technology? The answer is partly in the sheer complexity and unpredictability of the variables at play. While tech operates within the relatively defined boundaries of physics and engineering—where the art of the possible is constrained only by the laws of nature, capital, and the current state of technical prowess—the task of predicting how these innovations will weave into the fabric of society is infinitely more iffy.

Additionally, the science fiction audience may limit its influence on societal adoption. Books like *Snow Crash* have undeniably served as muses for the architects of the Metaverse. However, perhaps those at the helm of society's nontechnical spheres—policymakers, legislators, and business leaders—might not be as immersed in these stories. The technocrats laboring to construct the Metaverse might be well-versed in its science-fiction lore, but those responsible for guiding its assimilation into society could be navigating with less foresight.

Where does this leave the future of the Metaverse? Which sci-fi future lies ahead? The answer lies in twin truths: If we ignore dystopian predictions, they will be more likely to come true. Conversely, if we ignore optimistic predictions, they'll be less likely to arise. For the Metaverse to manifest more light than darkness, the insights of science fiction must reach those in positions of power far beyond the world of technology. In other words, the future of the Metaverse will be determined by how many of us are paying attention to it.

How can we educate and protect our kids as they grow up in an increasingly Meta world?

In 2023, a bipartisan group of 33 attorneys general from across the United States sued Meta for its alleged efforts to hook kids and teens on its products.[2] This ongoing case alleges Meta designed algorithms, notifications, and infinite scroll features that were specifically geared toward hooking kids on their product in violation of the Children's Online Privacy Protection Act (COPPA).[3] Whatever the legal veracity of these allegations, they serve to illustrate the possibility of using new, even more immersive, technologies like the Metaverse to hook kids on an array of digital products and services.

Broadly speaking, Metaverse platforms should be following the same basic ethical principles when it comes to advertising to kids as any other company, including being transparent, clearly labeling ads as such, not collecting children's personal identifiable information (PII) without parental consent (and likely not even then), and making privacy policies easily understandable. According to the Advertising Guidelines from the Children's Advertising Review Unit (CARU), "Children have limited knowledge, experience, sophistication, and maturity. Advertisers should recognize that younger children have a limited capacity to evaluate the credibility of information, may not understand the persuasive intent of advertising, and may not even understand that they are viewing or hearing advertising."[4] Some Metaverse platforms—including Meta itself—are already offering kids accounts that a parent or guardian can create for children aged 10–12; in 2023, Meta lowered this age group further. The consequences of such accessibility for children are yet to be determined, as it remains unclear how prolonged exposure to VR could affect a preteen's development.[5]

In general, parents may want to exercise caution and have frequent conversations with their children if they do allow them to use the Metaverse, even with kids' settings enabled. It is possible to further limit access to specific applications and disable further downloads without explicit parental permission. Having a clear playing space free of obstructions is also important to avoid injury. Plus, just as people (including kids) can suffer from motion sickness, the same thing can happen in the Metaverse— as can eye strain—so play time should be limited.[6] Further, just as some parents find it important to monitor their children's social media use in general, they may find equal importance in exercising caution with any interactions that their children may have in Metaverse environments, particularly with unknown users. Parents, for example, could prevent their kids from being searchable or contacted by strangers. Unfortunately, instances of cyberbullying, hate speech, and harassment have taken place in Metaverse environments with avatars exhibiting lewd behavior, seemingly, without consequence.[7] Other incidents involving sexual harassment have also been reported,[8] whereas there have also been examples of virtual avatars reporting rape in Metaverse settings.[9] Parents can block users and report abusive behavior to platforms like Meta or the FBI, though there is not yet the equivalent of a 911 for the Metaverse.[10]

Regardless of the final form the ecosystem takes, the Metaverse likely will be an important platform for a range of activity, and in the years ahead, it's likely that the first such experiences of users will be both early and formative. That is why, in an ideal scenario, the Metaverse would become something that social media promised but never attained—a platform for connection, growth, and even a resource for digital citizenship. If current trends continue, it's unlikely that the US federal government will be writing the rules and regulations of the Metaverse, meaning that most likely they will be coming from a combination of proactive US states and the European Union. Communciation and coordination will be vital in such a fragmented regulatory landscape. But what can users, who may not wish to let that process unfold, do? Will it be possible to "unplug" from the Metaverse?

What comes after the Metaverse?

Crystal balls are famously opaque. When the Pew Research Center asked an array of experts to predict the online environment of 2035, there was a range of predictions including the following: the seamless integration of digital and physical environments, the rise of public-spirited coders to promote digital literacy and citizenship, and the evolution of a digital sharing economy.[11] The Internet pioneer and Ostrom Workshop visiting scholar Doc Searls, for example, has a detailed vision based on ideas of self-sovereign identity, the Intention Byway, and so-called "palgorithms," which would permit users greater choice and transparency in the algorithms they use.[12]

It seems clear that cyberspace is here to stay—and, with it, different ways of experiencing both the virtual and augmented realities that it fuels. But will contact lens-equipped spatial computing, or neural-link chips enabling users to experience the world around them in new ways, exacerbate existing divides and problems or help us experience the world in new and exciting ways? Imagine walking down the street and being able to toggle the amount of information being shared about every individual, shop, plant, and animal you pass by. You could, in an instant, view the same street ten, fifty, or a hundred years ago, complete with changing the "costumes" of pedestrians around you. Similarly, imagine politicians being presented with information and statistics about the groups and audiences with which they interact, or business leaders and academics being able to pull an array of statistics, trend analyses, and witty one-liners up with a blink or gesture. Such a hyper-connected environment is, of course, ripe for abuse, further blurring reality and fiction, as it is with the possibility of civil and human rights protections.

Current trends fueling the rise, fall, and future of the Metaverse may abate or take on an entirely different form. Efforts to reign in, or at least label, AI-generated bots, deepfakes, and disinformation online could be largely successful. New tools to build consensus, protect users, and provide for extensive privacy protections, including biometric information, perhaps enshrined in a Digital Bill of Rights could be made widely available (and enabled by default).[13] Defining, to say nothing of operationalizing, such a cyber peace in an Internet of Things is no simple feat and is likely the work of generations. But, as the anthropologist Margaret Mead may have famously said, "Never doubt that a small group of thoughtful committed individuals can change the world; indeed, it's the only thing that ever has."

Conclusion

The Metaverse is a realm of infinite possibilities that, in many ways, is also a reflection of the best and worst of human nature. It is in many ways the ultimate manifestation, and potential realization, of William Gibson's original vision for cyberspace, one with the power to shape the economic, political, social, and legal trends of the twenty-first century. Or, it could be a fad, yet another repackaging of concepts and snake oil that have suffused the community since before the heady dot-com boom of the 1990s. Yet as seen during the COVID-19 pandemic, there is no substitute for face-to-face human interaction, regardless of how hyper-real an avatar might become. This was a theme of *Ready Player One*, in which the OASIS was shut down on certain days to promote a more balanced lifestyle. So far, though, average screen time per day shows no signs of abating on its own, increasing more than 50 minutes in the decade from 2013 to 2023 to more than seven hours per day.[14]

The future of the Metaverse need not be a dystopian matrix but neither is it likely to be a version of the holodeck with comprehensive safety features that can be easily turned on and off at a user's command. Navigating the myriad shades of gray between these extremes will no doubt be treacherous—platform owners, regulators, civil society groups, and users may get the balance wrong more than we get it right. But, just as no nation is an island in cyberspace, no persistent, immersive, virtual community is truly isolated; the ripples of regulatory actions, protests, and related technological tools from AI to quantum computing will shape our increasingly mixed reality for decades to come. Proactive engagement is key, including by using the tools presented by the Metaverse itself to organize across sectors and borders. The green shoots of such community activism have already begun to take root, but to be self-sustaining requires engaging users in cultivating and enforcing rules of behavior. As the Nobel Laureate Elinor (Lin) Ostrom taught us, effective communication is vital in this regard, along with efficient conflict resolution, opportunities for norm building, and enforcement of collective rules.[15] As Ostrom said, "Trust is the most important resource,"[16] and that's equally true in the Metaverse as it is in the real world.

NOTES

Prelims
1. Eric Ravenscraft, "What Is the Metaverse, Exactly?," *Wired* (Nov. 25, 2021), https://www.wired.com/story/what-is-the-metaverse/.
2. William Gibson, *Neuromancer* (2000), 6.
3. Ibid.
4. Frank Holmes, "The Metaverse Is A $1 Trillion Revenue Opportunity. Here's How to Invest...," *Forbes* (Dec. 20, 2021), https://www.forbes.com/sites/greatspeculations/2021/12/20/the-metaverse-is-a-1-trillion-revenue-opportunity-heres-how-to-invest/?sh=3bdcf2124df9.
5. Paul Solman, "Conventional Economists Sound Alarm over Cryptocurrency's Volatility," *PBS* (Apr. 8, 2022), https://www.pbs.org/newshour/show/conventional-economists-sound-alarm-over-cryptocurrencys-volatility.
6. "What is Metaverse Nation?," https://readyformetaverse.com/Unitedmetaversenation (last visited June 1, 2022).

Chapter 1
1. Bernard Marr, "The 10 Best Metaverse Quotes Everyone Should Read," *Forbes* (Apr. 15, 2022), https://www.forbes.com/sites/bernardmarr/2022/08/15/the-10-best-metaverse-quotes-everyone-should-read/?sh=491bee40225d.
2. Laura Dobberstein, "China Names Members of a Second Tech Supergroup to Define the Metaverse," *Register* (Jan. 23, 2024), https://www.theregister.com/2024/01/23/china_metaverse_standards_group/; Nataly Antonenko, "Chinese State Newspaper Develops NFT Platform Based on Metaverse," https://coinspaidmedia.com/news/china-daily-creates-metaverse-nft-marketplace/ (last visited Feb. 8, 2024).
3. Eric Ravenscraft, "What Is the Metaverse, Exactly?," *Wired* (Nov. 25, 2021), https://www.wired.com/story/what-is-the-metaverse/.
4. William Gibson, *Neuromancer* (2000), 6.
5. "cyberspace" *Computer Security Resource Center*, NIST. https://csrc.nist.gov/glossary/term/cyberspace (last visited Jan. 31, 2024).
6. Neal Stephenson, *Snow Crash* (1992), 102.
7. Matthew Ball, *The Metaverse: And How It Will Revolutionize Everything* (2022).

8. Stephenson, *supra* note 6, at 102.
9. Lik-Hang Lee et al., "What is the Metaverse? An Immersive Cyberspace and Open Challenges," ResearchGate (2022), https://www.researchgate.net/publication/361161855_What_is_the_Metaverse_An_Immersive_Cyberspace_and_Open_Challenges.
10. *See, for example*, Scott J. Shackelford, *The Internet of Things: What Everyone Needs to Know* (2020).
11. Frank Holmes, "The Metaverse Is A $1 Trillion Revenue Opportunity. Here's How to Invest...," *Forbes* (Dec. 20, 2021), https://www.forbes.com/sites/greatspeculations/2021/12/20/the-metaverse-is-a-1-trillion-revenue-opportunity-heres-how-to-invest/?sh=3bdcf2124df9.
12. Paul Solman & Lee Koromvokis, "Conventional Economists Sound Alarm over Cryptocurrency's Volatility," *PBS* (Apr. 8, 2022), https://www.pbs.org/newshour/show/conventional-economists-sound-alarm-over-cryptocurrencys-volatility.
13. *What is Metaverse Nation?*, https://readyformetaverse.com/Unitedmetaversenation#:~:text=What%20is%20United%20Metaverse%20Nation,are%20transparent%20and%20fully%20public (last visited June 1, 2022).
14. *See* Garry Williams, "The Metaverse Bubble is Bursting. Now the Real Work Begins," *Campaign* (Oct. 25, 2022), https://www.campaignasia.com/article/the-metaverse-bubble-is-bursting-now-the-real-work-begins/481920
15. *See* Dade Hayes, "Disney Unplugs Metaverse Unit During Initial Round of Layoffs," *Deadline* (Mar. 28, 2023), https://deadline.com/2023/03/disney-unplugs-metaverse-unit-layoffs-bob-chapek-bob-iger-1235311536/.
16. *See* Hannah Ritchie et al., "Internet," Our World in Data https://ourworldindata.org/internet (last visited Apr. 11, 2023).
17. Ibid.
18. *See* "A Brief Report on the World's Largest Metaverse Event 'Virtual Market 2022 Winter,'" Virtual Market (Feb. 6, 2023), https://virtual-market.prowly.com/226031-a-brief-report-on-the-worlds-largest-metaverse-event-virtual-market-2022-winter#:~:text=Market%202022%20Winter%20"-,A%20Brief%20Report%20on%20The%20World%27s,Event%20"Virtual%20Market%202022%20Winter%20"&text=HIKKY%20(Headquarters%3A%20Shibuya%2Dku,3%20to%2018%20December%202022.
19. Nick Bilton, "Why Tim Cook Is Going All In on the Apple Vision Pro," *Vanity Fair* (Feb. 1, 2024), https://www.vanityfair.com/news/tim-cook-apple-vision-pro.
20. Jay Yarow, "Paul Krugman Responds to All the People Throwing Around His Old Internet Quote," *Business Insider* (Dec. 30, 2013), https://www.businessinsider.com/paul-krugman-responds-to-internet-quote-2013-12.
21. Jordan Novet, "Mark Zuckerberg Envisions a Billion People in the Metaverse Spending Hundreds of Dollars Each," *CNBC* (June 22, 2022), https://www.cnbc.com/2022/06/22/mark-zuckerberg-envisions-1-billion-people-in-the-metaverse.html.
22. Mark Minevich, "The Metaverse and Web3 Creating Value in The Future Digital Economy," *Forbes* (June 17, 2022), https://www.forbes.com/sites/markminevich/2022/06/17/the-metaverse-and-web3-creating-value-in-the-future-digital-economy/?sh=73283fa17785.

23. *See* Janna Anderson & Lee Rainie, "A Sampling of Overarching Views on the Metaverse," Pew Research Center (Jun. 30, 2022), https://www.pewresearch.org/internet/2022/06/30/1-a-sampling-of-overarching-views-on-the-metaverse/.
24. Faye Angela Tricia Lopez, "35 Metaverse Quotes on Success" (Dec. 15, 2023), https://www.awakenthegreatnesswithin.com/35-metaverse-quotes-on-success/.
25. Justin Charity, "2022 Was the Year of the Metaverse—Until It Wasn't," Ringer (Dec. 29, 2022), https://www.theringer.com/tech/2022/12/29/23529842/metaverse-meaning-facebook-mark-zuckerberg-virtual-reality-game.
26. Ibid.
27. "The Best Metaverse Quotes," https://www.supplychaintoday.com/best-metaverse-quotes/ (last visited Feb. 21, 2024).
28. Soyeon Kim & Eunjoo Kim, "Emergence of the Metaverse and Psychiatric Concerns in Children and Adolescents," *Journal of Child and Adolescent Psychiatry* 34 (2023): 215.
29. Ibid.
30. "The Best Metaverse Quotes," *supra* note 27.
31. Ed Zitron, "RIP Metaverse," *Business Insider* (May 8, 2023), https://www.businessinsider.com/metaverse-dead-obituary-facebook-mark-zuckerberg-tech-fad-ai-chatgpt-2023-5.
32. Ibid.
33. Gene Marks, "This Week In Small Business Tech: Is The Metaverse Failing?" *Forbes* (Apr. 2, 2023), https://www.forbes.com/sites/quickerbettertech/2023/04/02/this-week-in-small-business-tech-is-the-metaverse-failing/?sh=24160d777a0d.
34. Ben Lang, "Meta Has Sold Nearly 20 Million Quest Headsets, But Retention Struggles Remain," *RoadtoVR* (Mar. 1, 2023, 6:36 AM), https://www.roadtovr.com/quest-sales-20-million-retention-struggles/.
35. Zitron, *supra* note 31.
36. *See, for example*, Tom Mitchelhill, "The Metaverse is Real: Zuck's 'Incredible' Photorealistic Tech Wows Crypto Twitter," *CoinTelegraph* (Sep. 29, 2023), https://cointelegraph.com/news/metaverse-podcast-lex-fridman-mark-zuckerberg-avatars-meta-codec.
37. Zitron, *supra* note 31.
38. Ibid.
39. Ibid.
40. Ibid.
41. Ibid.
42. Marks, *supra* note 33.
43. *See* Tripp Mickle & Brian X. Chen, "At Apple, Rare Dissent Over a New Product: Interactive Goggles," *New York Times* (Mar. 26, 2023), https://www.nytimes.com/2023/03/26/technology/apple-augmented-reality-dissent.html.

Chapter 2

1. Mathew Ball, *The Metaverse and How It Will Revolutionize Everything* (2022), 3.
2. Morgan Haefner & Ananya Bhattacharya, "The Metaverse Economy: Is It Real?" *Quartz* (Dec. 1, 2023), https://qz.com/emails/quartz-obsession/1851061482/the-metaverse-economy-is-it-real.

3. Michael R. Baye & Jeff T. Prince, *Managerial Economics and Business Strategy*, 10th ed. (2022), 3.
4. Steven A. Greenlaw & David Shapiro, *Principles of Microeconomics*, 2nd ed. (2011), 27.
5. Deven R. Desai & Mark A. Lemley, "Editorial: Scarcity, Regulation, and the Abundance Society," *Frontiers in Research Metrics & Analysis* 7, no. 1 (2023).
6. *See* Betsy Reed, "Man Who Paid $2.9m for NFT of Jack Dorsey's First Tweet Set to Lose almost $2.9m," *Guardian* (Apr. 14, 2022), https://www.theguardian.com/technology/2022/apr/14/twitter-nft-jack-dorsey-sina-estavi. (Showing an NFT depicting Jack Dorsey's first post on Twitter (now X) sold for $2.9 million, and when offered for sale one year later, had a top bid of less than $300.)
7. Lau Christensen & Alex Robinson, "The Potential Global Economic Impact of the Metaverse," (study, Analysis Group 2022), 1, https://www.analysisgroup.com/globalassets/insights/publishing/2022-the-potential-global-economic-impact-of-the-metaverse.pdf.
8. Ibid.
9. Andrew R. Chow, "Why TIME Is Launching a New Newsletter on the Metaverse," *TIME* (Nov. 18, 2021), https://time.com/6118513/into-the-metaverse-time-newsletter/ (last visited July 26, 2023).
10. Anton Korinek, "Metaverse Economics Part 1: Creating Value in the Metaverse," Brookings (June 6, 2023), https://www.brookings.edu/articles/metaverse-economics-part-1-creating-value-in-the-metaverse/.
11. Ibid.
12. Ibid.
13. Chow, *supra* note 9.
14. John Herrman & Kellen Browning, "Are We in the Metaverse Yet?" *New York Times* (Oct. 29, 2021), https://www.nytimes.com/2021/07/10/style/metaverse-virtual-worlds.html/.
15. Korinek, *surpa* note 10.
16. "Value Creation in the Metaverse," (report, McKinsey & Company, 2022), 29. https://www.mckinsey.com/%7E;/media/mckinsey/business%20functions/marketing%20and%20sales/our%20insights/value%20creation%20in%20the%20metaverse/Value-creation-in-the-metaverse.pdf.
17. Baye & Prince, *supra* note 3.
18. Herrman & Browning, *supra* note 14.
19. Ibid.; *see also* McKinsey & Company, *supra* note 16, at 43.
20. *See* McKinsey & Company, *supra* note 16, at 19.
21. *See* McKinsey & Company, *supra* note 16, at 30.
22. *See* Adnan Kayyali, "How Many Metaverses are There?" *Inside Telecom* (Sep. 9, 2022), https://insidetelecom.com/how-many-metaverses-are-there/.
23. Esther Shein, "Top Metaverse Platforms to Know About in 2023," *Tech Target* (Mar. 3, 2023), https://www.techtarget.com/searchcio/tip/Top-metaverse-platforms-to-know-about.
24. Ibid.
25. Ali Naheed, "EHR Interoperability Challenges and Solutions," EHR in Practice (Dec. 06, 2022), https://www.ehrinpractice.com/ehr-interoperability-challenges-solutions.html.

26. Tim Baharin, "The Four Major Players Battling To Own The Metaverse OS," *Forbes* (Nov. 18, 2022), https://www.forbes.com/sites/timbajarin/2022/11/18/the-four-major-players-battling-to-own-the-metaverse-os/?sh=3c84ba041e60.
27. "What Is Data Synchronization and Why Is It Important?" Talend. https://www.talend.com/resources/what-is-data-synchronization/, (last viewed Oct. 25, 2024).
28. Jyoti Mann, "Meta Has Spent $36 Billion Building the Metaverse but Still Has Little to Show for It, While Tech Sensations Such as the iPhone, Xbox, and Amazon Echo Cost Way Less," *Business Insider* (Oct. 29, 2022), https://www.businessinsider.com/meta-lost-30-billion-on-metaverse-rivals-spent-far-less-2022-10.
29. *See* Theo Priestley, "The Metaverse Needs An Operating System," *Medium* (Oct. 3, 2022), https://medium.com/@theo/the-metaverse-needs-an-operating-system-5fe135ec5b9.
30. Florian Buchholz et al., "There's More than One Metaverse," i-com (Nov. 16, 2022), https://www.degruyter.com/document/doi/10.1515/icom-2022-0034/html?lang=en.
31. John Sutton, *Sunk Costs and Market Structure* (Cambridge, MA: MIT Press, 2007), 3.
32. Greenlaw & Shapiro, *supra* note 4, at 216–217.
33. Michel Kilzi, "The New Virtual Economy of The Metaverse," *Forbes*, (May 20, 2022), https://www.forbes.com/sites/forbesbusinesscouncil/2022/05/20/the-new-virtual-economy-of-the-metaverse/?sh=3db3752f46d8.
34. McKinsey & Company, *supra* note 16, at 35.
35. Ibid., at 12.
36. Kilzi, *supra* note 33.
37. ABI Research, "Lower Entry Barrier and Improved Content Performance Create a 2024 Inflection Point for a US$100 Billion Virtual Reality Market by 2027," *PR Newswire* (Mar. 02, 2023), https://www.prnewswire.com/news-releases/lower-entry-barrier-and-improved-content-performance-create-a-2024-inflection-point-for-a-us100-billion-virtual-reality-market-by-2027-301760511.html.
38. Sean Middleton, "Five Ways Digital Technologies are Lowering Barriers to Entry," Cognizant (Nov. 12, 2015), https://www.cognizant.com/five-ways-digital-technologies-are-lowering-barriers-to-entry.
39. Jan Hatzius et al., "US Economics Analyst. Concentration, Competition, and the Antitrust Policy Outlook (Briggs/Phillips)," Goldman Sachs (July 18, 2021), https://www.gspublishing.com/content/research/en/reports/2021/07/19/ce786051-e0cd-46ba-8923-e30fc3673e9f.html.
40. Kasey Lobaugh, Bobby Stephens, & Jeff Simpson, "The Consumer is Changing, but Perhaps Not How You Think: A Swirl of Economic and Marketplace Dynamics is Influencing Consumer Behavior," Deloitte Insights (May 29, 2019), https://www2.deloitte.com/us/en/insights/industry/retail-distribution/the-consumer-is-changing.html.
41. Baye & Prince, *supra* note 3, at 44.
42. Bernard Marr, "The Metaverse and Digital Transformation At McDonald's," *Forbes* (June 22, 2022), https://www.forbes.com/sites/bernardmarr/2022/06/22/the-metaverse-and-digital-transformation-at-mcdonalds/?sh=20d86f203967.
43. Jelisa Castrodale, "Get Ready to Order Your Big Mac from a Virtual McDonald's in the Metaverse," *Food&Wine* (Feb.11, 2022), https://www.foodandwine.com/news/mcdonalds-metaverse-restaurants-ordering.

44. McKinsey & Company, *supra* note 16.
45. Jessica Golden, "Nike Teams up with Roblox to Create a Virtual World called Nikeland," *CNBC* (Nov. 19, 2021), https://www.cnbc.com/2021/11/18/nike-teams-up-with-roblox-to-create-a-virtual-world-called-nikeland-.html.
46. Kathy Hirsh-Pasek et al., "A Whole New World: Education Meets the Metaverse," Brookings (Feb. 14, 2022), https://www.brookings.edu/articles/a-whole-new-world-education-meets-the-metaverse/.
47. Baye & Prince, *supra* note 3.
48. Austan Goolsbee, "Competition in the Computer Industry: Online Versus Retail," *Journal of Industrial Economics* 49, no. 4 (2001): 487–499.
49. Jeffrey Prince, "The Beginning of Online/Retail Competition and Its Origins," *International Journal of Industrial Organization* 25, no. 1 (2007): 139–156.
50. Korinek, *supra* note 10.
51. Bain & Company, "Young Gamers are Embracing the Metaverse," *PR Newswire* (July 26, 2022), https://www.prnewswire.com/news-releases/young-gamers-are-embracing-the-metaverse-301593398.html.
52. Talespin Team, "Why Virtual Reality is Important: Exploring the Benefits of VR," Talespin (July 10, 2023), https://www.talespin.com/reading/why-virtual-reality-is-important-exploring-the-benefits-of-vr.
53. Volodymyr Fedorychak, "How AR/VR Can Help to Create Product Demos and Increase Sales?" SmartTek Solutions (Sep.18, 2023), https://smarttek.solutions/blog/how-ar-vr-help-to-product-demos/.
54. Tomislav Bezmalinovic, "We are one Step Closer to a Google Earth VR for standalone headsets," Mixed (May 27, 2023), https://mixed-news.com/en/photorealistic-3d-tiles-standalone-google-earth-vr/.
55. Scott Stein & Imad Khan, "Your Google Maps Experience Is About to Get More Immersive," *CNET* (May 11, 2023), https://www.cnet.com/tech/computing/google-builds-ar-for-everywhere-from-maps/.
56. Mandy Erikson, "Virtual Reality System Helps Surgeons, Reassures Patients," Stanford Medicine (2017), https://med.stanford.edu/news/all-news/2017/07/virtual-reality-system-helps-surgeons-reassures-patients.html.

Chapter 3

1. "18 Freedom of Speech Quotes You Should Know," Freedom Forum, https://www.freedomforum.org/freedom-of-speech-quotes/ (last visited Feb. 12, 2024).
2. M. I. Berkman, & E. Akan, "Presence and Immersion in Virtual Reality," in *Encyclopedia of Computer Graphics and Games*, ed. N. Lee, (Cham: Springer, 2024), 1461–1470, https://doi.org/10.1007/978-3-319-08234-9_162-1.
3. R. Warp et al., "Validating the Effects of Immersion and Spatial Audio Using Novel Continuous Biometric Sensor Measures for Virtual Reality," *2022 IEEE International Symposium on Mixed and Augmented Reality Adjunct (ISMAR-Adjunct)* (2022): 262–265, https://www.semanticscholar.org/paper/453176799b7ebbf52133e1ec49aa27f553779d3f.
4. Olin, Patrick Aggergaard, et al., "Designing for Heterogeneous Cross-Device Collaboration and Social Interaction in Virtual Reality," *Proceedings of the 32nd Australian Conference on Human-Computer Interaction* (2020): n. p., https://www.

semanticscholar.org/paper/Designing-for-Heterogeneous-Cross-Device-and-Social-Olin-Issa/d68a88824a2543c0120f6f9fea9befc9ab6cac5b?utm_source=direct_link.
5. Wendy L. Patrick, "Sexual Assault in the Metaverse: Virtual Reality, Real Trauma," *Psychology Today* (Jan. 3, 2023), https://www.psychologytoday.com/us/blog/why-bad-looks-good/202301/sexual-assault-in-the-Metaverse-virtual-reality-real-trauma.
6. Adam Smith, "Meta Forced to Add 'Personal Boundaries' to the Metaverse After Woman Was Sexually Harassed in Virtual Reality," *Independent* (Feb. 4, 2022), https://www.independent.co.uk/tech/meta-personal-boundaries-Metaverse-sexual-harass-b2007878.html.
7. Althoff-Thomson Savannah & Jean-Paul Van Belle, "The Metaverse: Investigating the Motivations and Experiences of Early Adopters in RecRoom," *18th Iberian Conference on Information Systems and Technologies* (2023): 1–6 https://doi.org/10.23919/CISTI58278.2023.10211253.
8. Margaret E. Morris, et al., "'I Don't Want to Hide Behind an Avatar': Self-Representation in Social VR Among Women in Midlife," *Proceedings of the 2023 ACM Designing Interactive Systems Conference* (2023), https://dl.acm.org/doi/pdf/10.1145/3563657.3596129.
9. Yubo Kou & Xinning Gui, "Harmful Design in the Metaverse and How to Mitigate It: A Case Study of User-Generated Virtual Worlds on Roblox," *Proceedings of the 2023 ACM Designing Interactive Systems Conference* (2023), https://doi.org/10.1145/3563657.3595960.
10. Louis B. Rosenberg, "The Metaverse and Conversational AI as a Threat Vector for Targeted Influence," *IEEE 13th Annual Computing and Communication Workshop and Conference (CCWC)* (2023), https://doi.org/10.1109/CCWC57344.2023.10099167.
11. Ibid.
12. Ibid.
13. Hongni Ye, "Twilight Rohingya: The Design and Evaluation of Different Navigation Controls in a Refugee VR Environment," *International Conference on Cyber Worlds* (Sep. 1, 2022), https://www.semanticscholar.org/paper/04bf70b3912ea01224c71473d12f6b03ea8b7115.
14. N. Van Veelen et al., "Tailored Immersion: Implementing Personalized Components Into Virtual Reality for Veterans With Post-Traumatic Stress Disorder," *European Psychiatry* (2022), https://www.ncbi.nlm.nih.gov/pmc/articles/PMC9567913/.
15. "LookBack: Virtual Therapy for Dementia," Virtue Health, https://www.virtue.io/lookback/ (last visited July 24, 2024); Lucy Johnson, "The Future of Dementia Care Is Virtual Reality," *Wired* (Mar. 23, 2024), https://www.wired.com/story/virtual-reality-dementia-technology/; "The Wayback VR," The Wayback, https://thewaybackvr.com/ (last visited July 24, 2024).
16. Kathy Hirsh-Pasek et al., "A Whole New World: Education Meets the Metaverse," Brookings (Feb. 14, 2022), https://www.brookings.edu/articles/a-whole-new-world-education-meets-the-Metaverse/.
17. "Education in the Metaverse," YouTube, https://www.youtube.com/watch?v=KLOcj5qvOio; https://www.youtube.com/watch?v=80IIEnSNwQc.
18. L. Procter, "I Am/We Are: Exploring the Online Self-Avatar Relationship," *Journal of Communication Inquiry* 45, no. 1 (2022): 45–64 https://doi.org/10.1177/0196859920961041.

19. Jia Wei, "Self-Representation through Online Avatars in Major Online Contexts Among Chinese University Students," *Journal of Education, Humanities, & Social Sciences* (2023), https://drpress.org/ojs/index.php/EHSS/article/download/4674/4525; Jia Wei, "Self-representation through Online Avatars in Major Online Contexts Among Chinese University Students," *Journal of Education, Humanities, & Social Sciences* (2023), https://www.semanticscholar.org/paper/72db151e135fea42fcfba542147be9c728875fba; Mila Bujić et al., "Self-Representation Does (Not) Spark Joy: Experiment on Effects of Avatar Customisation and Personality on Emotions in VR," *GamiFIN Conference* (2023), https://ceur-ws.org/Vol-3405/paper9.pdf.
20. L. Procter, "I Am/We Are: Exploring the Online Self-Avatar Relationship," *Journal of Communication Inquiry* 45, no. 1 (2021): 45–64, https://doi.org/10.1177/0196859920961041.
21. Daniel Zimmermann et al., "Self-Representation through Avatars in Digital Environments," *Current Psychology* 42 (2022): 21775–21789, https://link.springer.com/content/pdf/10.1007/s12144-022-03232-6.pdf.
22. Zimmermann, *supra* note 19; Morris, *supra* note 8; Bujić et al., *supra* note 17.
23. Panote Siriaraya & Chee Siang Ang, "The Social Interaction Experiences of Older People in a 3D Virtual Environment," *Human–Computer Interaction Series* (2019), https://www.semanticscholar.org/paper/The-Social-Interaction-Experiences-of-Older-People-Siriaraya-Ang/c647b5404e65fc07a8343fc840868ea5cbfc7cac?utm_source=direct_link.
24. H. Zhao et al., "Hand-in-Hand: A Communication-Enhancement Collaborative Virtual Reality System for Promoting Social Interaction in Children with Autism Spectrum Disorders," *IEEE Transactions on Human-Machine Systems* 48, no. 2 (2018): 136–148, https://pubmed.ncbi.nlm.nih.gov/30345182/.
25. A. Lewinski et al., "Type 2 Diabetes Education and Support in a Virtual Environment: A Secondary Analysis of Synchronously Exchanged Social Interaction and Support," *Journal of Medical Internet Research* 20, no. 2 (2018): e61, https://www.ncbi.nlm.nih.gov/pmc/articles/PMC5842323/.
26. Ibid. Esubalew Bekele et al., "Multimodal adaptive social interaction in a virtual environment (MASI-VR) for children with Autism spectrum disorders (ASD)," *2016 IEEE Virtual Reality and 3D User Interfaces*, (2016): 121–130, https://www.semanticscholar.org/paper/Multimodal-adaptive-social-interaction-in-virtual-Bekele-Wade/2527f981c8c51ddba92116a8bd7f54324067dc18?utm_source=direct_link; Zhao, *supra* note 22.
27. Yixuan Zhang, "A Study on the Para-social Interaction Between Idols and Fans in Virtual Applications," *Advances in Social Science, Education and Humanities Research* (2022), https://www.semanticscholar.org/paper/A-Study-on-the-Para-social-Interaction-Between-and-Zhang/a3d72b70bb5203cd7cdaa07282a78f3a6fd5017e?utm_source=direct_link.
28. Anna Codrea-Rado, "Virtual Vandalism: Jeff Koons's 'Balloon Dog' Is Graffiti-Bombed," *New York Times* (Oct. 10, 2017), https://www.nytimes.com/2017/10/10/arts/design/augmented-reality-jeff-koons.html; Geraint Lloyd-Taylor, "Virtual Vandalism? Taking the Art of Protest to Another Dimension," Lewis Silkin (Oct. 9, 2017), https://adlaw.lewissilkin.com/post/102ehec/virtual-vandalism-taking-the-art-of-protest-to-another-dimension.

29. Siobhan Hanna, "Why content moderation could make or break the Metaverse," *Fast Company* (Nov. 16, 2022), https://www.fastcompany.com/90811476/why-content-moderation-could-make-or-break-the-metaverse.
30. Ryan Hsu, "Meet the new 'verse, same as the old 'verse: Moderating the 'Metaverse,'" Georgetown Law Technology Review (May 2022), https://georgetownlawtechreview.org/meet-the-new-verse-same-as-the-old-verse-moderating-the-Metaverse/GLTR-05-2022/.
31. Katie Paul, "Meta Releases AI Model for Translating Speech Between Dozens of Languages," *Reuters* (Aug. 22, 2023), https://www.reuters.com/technology/meta-releases-ai-model-translating-speech-between-dozens-languages-2023-08-22/.
32. Nazanin Sabri et al., "Challenges of Moderating Social Virtual Reality," *Proceedings of the 2023 CHI Conference on Human Factors in Computing Systems* article 384 (2023): 1–20, https://doi.org/10.1145/3544548.3581329.
33. M. van Wegen et al., "An Overview of Wearable Haptic Technologies and Their Performance in Virtual Object Exploration," *Sensors (Basel)* 23, no. 3 (2023):1563. doi: 10.3390/s23031563. PMID: 36772603; PMCID: PMC9919508.
34. J. Seering et al., "Moderator Engagement and Community Development in the Age of Algorithms," *New Media & Society* 21 (2019): 1417.
35. Nazanin Sabri et al., "Challenges of Moderating Social Virtual Reality," *Proceedings of the 2023 CHI Conference on Human Factors in Computing Systems* article 384 (2023): 1–20, https://doi.org/10.1145/3544548.3581329.
36. Siobhan Hanna, "Why Content Moderation Could Make or Break the Metaverse," *Fast Company* (Nov. 16, 2022), https://www.fastcompany.com/90811476/why-content-moderation-could-make-or-break-the-metaverse; "OpenAI Says AI Tools Can Be Effective for Content Moderation," *Reuters* (Aug. 15, 2023), https://www.reuters.com/technology/openai-says-ai-tools-can-be-effective-content-moderation-2023-08-15/.
37. Hanna, *supra* note 29.
38. Mitchell Goldberg & Fabian Schär, "Metaverse Governance: An Empirical Analysis of Voting Within Decentralized Autonomous Organizations," *Journal of Business Research* 160 (2023): article 113764, https://doi.org/10.1016/j.jbusres.2023.113764; E. Karaarslan & S. Yazici, "Metaverse and Decentralization," *in* Metaverse (Studies in Big Data), eds. F. S. Esen, H. Tinmaz & M. Singh, (2023); Justin Sun, "Why the Future of the Metaverse Can Only Be Decentralized," *VentureBeat* (Mar. 5, 2022), https://venturebeat.com/datadecisionmakers/why-the-future-of-the-metaverse-can-only-be-decentralized/ (arguing that the future of the metaverse lies in decentralization and the transformative power of decentralized autonomous organizations (DAOs) over centralized models).
39. *See* Barry Solaiman, "Telehealth in the Metaverse: Legal & Ethical Challenges for Cross-Border Care in Virtual Worlds," *Journal of Law and Medical Ethics* 51 (2023): 287, 288.
40. 17 U.S.C. § 102 ("... authorship fixed in any tangible medium of expression ...").
41. Kathryn Park, "Trademarks in the Metaverse," *WIPO Magazine* (Mar. 2022), https://www.wipo.int/wipo_magazine/en/2022/01/article_0006.html.
42. The Lanham Act defines "use in commerce" as the use of a mark "on services when it is used or displayed in the sale or advertising of services and the services are rendered in commerce." 15 U.S.C. § 1127.

43. Rhys Thomas, "Miley Cyrus, Rihanna, Snoop Dogg: Meet the Celebrities of the Metaverse," *Face* (Feb. 23, 2024), https://theface.com/life/miley-cyrus-rihanna-snoop-dogg-metaverse-celebrities.
44. U.S. Constitutional Amendments I–X.
45. Eugene Volokh, "In Defense of the Marketplace of Ideas/Search for Truth as a Theory of Free Speech Protection," *Virginia Law Review* 97 (2011): 595.
46. U.S. Constitutional Amendment I.
47. "Crime and Punishment in the Metaverse: A Primer," Observer Research Foundation (Feb. 17, 2022), https://www.orfonline.org/research/crime-and-punishment-in-the-metaverse-a-primer
48. Liat Franco & Khalid Ghanayim, "The Criminalization of Cyberbullying Among Children and Youth," *Santa Clara Journal of International Law* 17 no. 287 (2019).
49. Ibid.
50. Children's Online Privacy Protection Act of 1998, 15 U.S.C. §§ 6501-6506 (2012).
51. Aaron Bryant, "The Effect of Social Media on the Physical, Social Emotional, and Cognitive Development of Adolescents," Merrimack Scholarworks (2018), https://scholarworks.merrimack.edu/honors_capstones/37/.

Chapter 4

1. "Hiscox Cyber Readiness Report 2021," Hiscox (2021), https://www.hiscoxgroup.com/sites/group/files/documents/2021-04/Hiscox%20Cyber%20Readiness%20Report%202021.pdf (last visited Jan. 4, 2024).
2. Steve Zurier, "Average cost of a data breach expected to hit $5 million in 2023," *SC Media* (Dec. 19, 2022), https://www.scmagazine.com/news/email-security/average-cost-of-a-data-breach-expected-to-hit-5-million-in-2023.
3. Offering a comprehensive review of the types of actors, threats, and cyberattack costs for the year 2016, Council of Economic Advisers, "The Cost of Malicious Cyber Activity to the U.S. Economy," (Feb. 2018), https://trumpwhitehouse.archives.gov/wp-content/uploads/2018/02/The-Cost-of-Malicious-Cyber-Activity-to-the-U.S.-Economy.pdf.
4. Bharath Aiyer et al., "New Survey Reveals $2 Trillion Market Opportunity for Cybersecurity Technology and Service Providers," McKinsey & Company (Oct. 27, 2022), https://www.mckinsey.com/capabilities/risk-and-resilience/our-insights/cybersecurity/new-survey-reveals-2-trillion-dollar-market-opportunity-for-cybersecurity-technology-and-service-providers.
5. Bruce Schneier, *A Hacker's Mind: How the Powerful Bend Society's Rules, and How to Bend Them Back*, (New York: W. W. Norton, 2023), 2.
6. *See, for example,* Walter Isaacson, *The Innovators: How a Group of Hackers, Geniuses, and Geeks Created the Digital Revolution* (New York: Simon & Schuster, 2015), 203.
7. Ibid., at 202.
8. Ava Ex Machina, "Hacking the Holocaust" *Medium* (Sep. 22, 2017), https://medium.com/@silicondomme/hacking-the-holocaust-abcd332947ae.
9. *See* Elizabeth McCracken, "Dial-Tone Phreak," *New York Times* (Dec. 30, 2007), https://www.nytimes.com/2007/12/30/magazine/30joybubbles-t.html. For more on the history of hacks, *see* Scott J. Shackelford & Scott O. Bradner, *Forks in the Digital Road: Key Decisions that Gave us the Internet We Have* (New York: Oxford University Press, 2024).

10. *See* Dakota Murphey, "A History of Information Security," *IFSec Global* (June 27, 2019), https://www.ifsecglobal.com/cyber-security/a-history-of-information-security/.
11. *See* Schneier, *supra* note 5, at 12.
12. *See* "Five Interesting Facts About the Morris Worm (for its 25th Anniversary)," *WeLiveSecurity* (Nov. 6, 2013), https://www.welivesecurity.com/2013/11/06/five-interesting-facts-about-the-morris-worm-for-its-25th-anniversary/.
13. Worms and viruses are similar, but different in one key way: A virus needs an external command, from a user or a hacker, to run its program. A worm, by contrast, hits the ground running all on its own.
14. "The Morris Worm: The First Significant Cyber Attack," ThinkReliability, https://www.thinkreliability.com/InstructorBlogs/blog-MorrisWorm.pdf (last visited Jan. 9, 2024).
15. Timothy B. Lee, "How a Grad Student Trying to Build the First Botnet Brought the Internet to Its Knees," *Washington Post* (Nov. 1, 2013), https://www.washingtonpost.com/news/the-switch/wp/2013/11/01/how-a-grad-student-trying-to-build-the-first-botnet-brought-the-internet-to-its-knees/.
16. Andrew W. Murray, *The Regulation of Cyberspace: Control in the Online Environment* (New York: Routledge, 2007), 44.
17. *Viruses/Contaminant/Destructive Transmission Statutes*, Nat'l Conf. St. Legislatures, http://www.ncsl.org/IssuesResearch/TelecommunicationsInformationTechnology/StateVirusandComputerContaminantLaws/tabid/13487/Default.aspx (last updated Feb. 14, 2012). For a deeper dive on this topic, see Chapter 3, Scott J. Shackelford, *Managing Cyber Attacks in International Law, Business and Relations: In Search of Cyber Peace* (New York: Cambridge University Press, 2014).
18. For more on the topic of defining "reasonable" cybersecurity, see Scott J. Shackelford, Anne Boustead & Christos Madrikis, "Defining 'Reasonable' Cybersecurity: Lessons from the States," *Yale Journal of Law and Technology* 25, no. 86 (2023).
19. This subsection was first published as Scott J. Shackelford, "Zero-Trust Security: Assume That Everyone and Everything on the Internet Is Out to Get You—and Maybe Already Has," Conversation (May 21, 2021), https://theconversation.com/zero-trust-security-assume-that-everyone-and-everything-on-the-internet-is-out-to-get-you-and-maybe-already-has-160969.
20. "Zero Trust Security: Multi-Layered Protection against Cyber-Threats," CipherSpace, https://www.cipherspace.com/infographics/zero-trust-security/ (last visited May 11, 2023).
21. *See, for example*, Chelsea Ong, "The Metaverse May Bring New Cyber Risks. Here's What Companies Can Do," *CNBC* (Mar. 22, 2022), https://www.cnbc.com/2022/03/23/the-metaverse-may-bring-new-cyber-risks-heres-what-firms-can-do.html.
22. *See* Kevin Granville, "Facebook and Cambridge Analytica: What You Need to Know as Fallout Widens," *New York Times* (Mar. 19, 2018), https://www.nytimes.com/2018/03/19/technology/facebook-cambridge-analytica-explained.html.
23. *See* Scott J. Shackelford, "Facebook's Social Responsibility Should Include Privacy Protection," Conversation (Apr. 12, 2018), https://theconversation.com/facebooks-social-responsibility-should-include-privacy-protection-94549. *See, for example*, Graison Dangor, "Facebook Overpaid FTC Fine by Billions To Protect Zuckerberg,

Lawsuits Say," *Forbes* (Sep. 21, 2021), http://www.forbes.com/sites/graisondangor/2021/09/21/facebook-overpaid-ftc-fine-by-billions-to-protect-zuckerberg-lawsuits-say/?sh=48c479f352e2.

24. *See* Jinwon Choi et al., "Analysis of Cybersecurity Threats and Vulnerabilities in Metaverse Environment, 융합보안논문지," *Journal of Convergence Security* 22, no. 3 (2022): 19–24.

25. *The Trust Machine, supra* note 19; Theodore Kinni, "Tech Savvy: How Blockchains Could Transform Management," *MIT Sloan Management Review* (May 12, 2016), http://sloanreview.mit.edu/article/tech-savvy-how-blockchains-could-transform-management/?utm_source=twitter&utm_medium=social&utm_campaign=smdirect ("Now imagine the opportunities that arise from the ability to search the World Wide Ledger, a decentralized database of much of the world's structured information. Who sold which discovery to whom? At what price? Who owns this intellectual property? Who is qualified to handle this project? What medical skills does our hospital have on staff? Who performed what type of surgery with what outcomes? How many carbon credits has this company saved? Which suppliers have experience in China? What subcontractors delivered on time and on budget according to their smart contracts? The results of these queries won't be resumes, advertising links, or other pushed content; they'll be transaction histories, proven track records of individuals and enterprises, ranked perhaps by reputation score."). For more on this topic, see Scott J. Shackelford & Steve Myers, "Block-by-Block: Leveraging the Power of Blockchain Technology to Build Trust and Promote Cyber Peace," *Yale Journal of Law and Technology* 19 (2017): 334.

26. *See* Bill Bowman, "How Blockchain Can Be Hacked: The 51% Rule and More," Security Boulevard (Feb. 27, 2019), https://securityboulevard.com/2019/02/how-blockchain-can-be-hacked-the-51-rule-and-more/.

27. Joao Marinotti, "Can You Truly Own Anything in the Metaverse? A Law Professor Explains How Blockchains and NFTs Don't Protect Virtual Property," Conversation (Apr. 21, 2022), https://theconversation.com/can-you-truly-own-anything-in-the-metaverse-a-law-professor-explains-how-blockchains-and-nfts-dont-protect-virtual-property-179067.

28. Yang-Wai Chow, "Visualization and Cybersecurity in the Metaverse: A Survey," *Journal of Imaging* 9 no. 1 (2023): 11, https://doi.org/10.3390/jimaging9010011.

29. *See, for example*, Glorin Sebastian, "A Descriptive Study on Metaverse: Cybersecurity Risks, Controls, and Regulatory Framework," IGI Global (2023), https://www.igi-global.com/article/a-descriptive-study-on-metaverse/315591; Edlyn V. Levine & Algirde Pipikaite, "Hardware is a cybersecurity risk. Here's what we need to know," World Economic Forum (Dec. 19, 2019), https://www.weforum.org/agenda/2019/12/our-hardware-is-under-cyberattack-heres-how-to-make-it-safe/.

30. *See* Kevin James, "Cybersecurity Risks in The Metaverse: A Complete Guide," CybersecurityForMe (2023), https://cybersecurityforme.com/cybersecurity-risks-in-the-metaverse/.

31. *See* Dylan Curran, "Are You Ready? Here is All the Data Facebook and Google Have on You," *Guardian* (Mar. 30, 2018), https://www.theguardian.com/commentisfree/2018/mar/28/all-the-data-facebook-google-has-on-you-privacy.

32. Ibid.

33. *See* Sydney Butler, "What Are Facebook Shadow Profiles, and Should You Be Worried?," *How-to-Geek* (Dec. 10, 2021), https://www.howtogeek.com/768652/what-are-facebook-shadow-profiles-and-should-you-be-worried/.
34. *See, for example,* Anokhy Desai, "Indiana Governor Signs a Comprehensive Privacy Act into Law," IAPP (May 3, 2023), https://iapp.org/news/a/indiana-governor-signs-a-comprehensive-privacy-act-into-law/.
35. Lee Rainie, "Americans' Complicated Feelings about Social Media in an Era of Privacy Concerns," Pew Research Center (Mar. 27, 2018), http://www.pewresearch.org/fact-tank/2018/03/27/americans-complicated-feelings-about-social-media-in-an-era-of-privacy-concerns/.
36. *See* Brooke Auxier et al., "Americans and Privacy: Concerned, Confused and Feeling Lack of Control Over Their Personal Information," Pew Research Center (Nov. 15, 2019), https://www.pewresearch.org/internet/2019/11/15/americans-and-privacy-concerned-confused-and-feeling-lack-of-control-over-their-personal-information/.
37. *See* Bree Fowler, "Your Digital Footprint: It's Bigger Than You Realize," *CNET* (Apr. 4, 2022), https://www.cnet.com/news/privacy/features/your-digital-footprint-its-bigger-than-you-realize/.
38. Sacha Molitorisz, "It's Time for Third-Party Data Brokers to Emerge from the Shadows," Conversation (Apr. 4, 2018), https://theconversation.com/its-time-for-third-party-data-brokers-to-emerge-from-the-shadows-94298.
39. Ibid.
40. Shobhit Seth, "How Much Can Facebook Potentially Make from Selling Your Data?," Investopedia (Apr. 11, 2018), https://www.investopedia.com/tech/how-much-can-facebook-potentially-make-selling-your-data/.
41. "What Are Data Brokers—And What Is Your Data Worth?," WebFX, https://www.webfx.com/blog/general/what-are-data-brokers-and-what-is-your-data-worth-infographic/ (last visited Mar. 23, 2019).
42. "Data Brokers Market Estimated to Reach US$ 462.4 billion by 2031," *TMR Report*, Transparency Market Res. (Aug. 1, 2022), https://www.globenewswire.com/news-release/2022/08/01/2489563/0/en/Data-Brokers-Market-Estimated-to-Reach-US-462-4-billion-by-2031-TMR-Report.html.
43. Ibid.
44. Ibid.
45. Ibid.
46. Ibid.
47. Ibid.
48. Ibid.
49. Ibid.
50. Mitra Pooyandeh et al., "Cybersecurity in the AI-Based Metaverse: A Survey," *Applied Sciences* 12, no. 24 (2022): 12993, https://www.mdpi.com/2076-3417/12/24/12993.
51. "Metaverse and Space May Be Cybercriminals' Playgrounds in 2023," Business Wire (Dec. 7, 2022), https://www.businesswire.com/news/home/20221207005227/en/Metaverse-and-Space-May-Be-Cybercriminals'-Playgrounds-in-2023.
52. *See* "Metaverse Security: Emerging Scams and Phishing Risks," PwC, https://www.pwc.com/us/en/tech-effect/cybersecurity/emerging-scams-and-phishing-risks-in-the-metaverse.html (last visited May 12, 2023).

53. "See Cyber Daily: Come the Metaverse, Can Privacy Exist?," Wall Street Journal (Jan. 4, 2022), https://www.wsj.com/articles/cyber-daily-come-the-metaverse-can-privacy-exist-11641304332.
54. See Jinwon Choi et al., "Analysis of Cybersecurity Threats and Vulnerabilities in Metaverse Environment," 융합 보안논 문지 Yunghabboannonmunji [Convergence Security Journal] (2022), https://www.earticle.net/Article/A419097.
55. Naomi Nix, "Attacks in the Metaverse are Booming. Police are Starting to Pay Attention," Washington Post (Feb. 4, 2024), https://www.washingtonpost.com/technology/2024/02/04/metaverse-sexual-assault-prosecution/.
56. See "Metaverse Security," supra note 52.
57. Ibid.
58. Ibid.
59. Ibid.
60. Ibid.
61. Ibid.
62. "How to Protect Yourself Against Crime in the Metaverse?: A Comprehensive Guide," ZebPay (Apr. 14, 2023), https://zebpay.com/blog/how-to-protect-yourself-from-crime-in-the-metaverse.
63. See "Multi-Factor Authentication," Cybersecurity and Infrastructure Security Agency, https://www.cisa.gov/sites/default/files/publications/MFA-Fact-Sheet-Jan22-508.pdf (last visited Nov. 9, 2023).
64. Ibid.
65. Jess Weatherbed, "Experts Link LastPass Security Breach to a String of Crypto Heists," Verge (Sep. 7, 2023), https://www.theverge.com/2023/9/7/23862658/lastpass-security-breach-crypto-heists-hackers.
66. See "4 Things You Can Do To Keep Yourself Cyber Safe," Cybersecurity and Infrastructure Security Agency, https://www.cisa.gov/news-events/news/4-things-you-can-do-keep-yourself-cyber-safe (last updated Dec. 18, 2022).
67. See Emil Sayegh, "How to Secure a Metaverse," Forbes (Apr. 28, 2022), https://www.forbes.com/sites/emilsayegh/2022/04/28/how-to-secure-a-metaverse/?sh=1590db9a26c3.
68. See Glorin Sebastian, "A Descriptive Study on Metaverse: Cybersecurity Risks, Controls, and Regulatory Framework," International Journal of Security and Privacy in Pervasive Computing 15 1 (2023).
69. See Shelby Brown, "The Data Privacy Tips Digital Security Experts Wish You Knew," CNET (May 3, 2022), https://www.cnet.com/tech/services-and-software/the-data-privacy-tips-digital-security-experts-wish-you-knew/.
70. See Ruben Merre, "Five Tips For Staying Safe In The Metaverse," Ngrave (Mar. 21, 2022), https://www.ngrave.io/en/blog/five-tips-for-staying-safe-in-the-metaverse.
71. See "Metaverse and Space May Be Cybercriminals' Playgrounds in 2023," Business Wire (Dec. 7, 2022), https://www.businesswire.com/news/home/20221207005227/en/Metaverse-and-Space-May-Be-Cybercriminals%E2%80%99-Playgrounds-in-2023.
72. See Chelsea Ong, "The Metaverse May Bring New Cyber Risks. Here's What Companies Can Do," CNBC (Mar. 22, 2022), https://www.cnbc.com/2022/03/23/the-metaverse-may-bring-new-cyber-risks-heres-what-firms-can-do.html.
73. Ibid.

74. *See* Lisa Sotto & Samuel Grogan, "Privacy and Cybersecurity Risks in the Metaverse: 5 Steps to Protect your Data," ALM | Property Casualty 360 (Apr. 27, 2023), https://www.propertycasualty360.com/2023/04/27/privacy-and-cybersecurity-risks-in-the-metaverse-5-steps-to-protect-your-data-414-237513/.
75. *See* "What is the Metaverse?," Avast, https://www.avast.com/c-metaverse (last visited Nov. 9, 2023).
76. *See* Sayegh, *supra* note 67.
77. *See* Ong, *supra* note 72.
78. *See* "What is Encryption?," Cloudflare, https://www.cloudflare.com/en-gb/learning/ssl/what-is-encryption/ (last visited Nov. 9, 2023).
79. Ibid.
80. *See* Choi et al., *supra* note 39.
81. *See* Zebpay, *supra* note 62.
82. *See* "How to Secure the Metaverse with 4 Key Tips," Quasa, https://quasa.io/media/how-to-secure-the-metaverse-with-4-key-tips (last visited Nov. 9, 2023).
83. *See* Zebpay, *supra* note 62.
84. *See* Yang-Wai Chow et al., "Visualization and Cybersecurity in the Metaverse: A Survey," *Journal of Imaging* 9 (2023): 11.
85. *See* Zebpay, *supra* note 62.
86. *See* Merre, *supra* note 70.
87. Nix, *supra* note 55.
88. *See* Brenda K. Wiederhold, "Sexual Harassment in the Metaverse ," *Cyberpsychology, Behavior, and Social Networking* 25 (2022): 479.
89. Ibid.
90. *See* Sayegh, *supra* note 67.
91. *See* Cathy Li & Farah Lalani, "How to Address Digital Safety in the Metaverse," World Economic Forum (Jan. 14, 2022), https://www.weforum.org/agenda/2022/01/metaverse-risks-challenges-digital-safety/.
92. *See* Mark Read, "We Need to Make the Metaverse Safe for Everyone. Here's How," World Economic Forum (May 22, 2022), https://www.weforum.org/agenda/2022/05/heres-how-to-make-the-metaverse-safe/.
93. *See* Doina Banciu et al., "Cyber Security and Human Rights Considering the Metaverse," *Jurnalul Libertății de Conștiință [Journal for Freedom of Conscience]* 9 (2021): 638.
94. *See* Vaishnavi Joshi, "India: Importance of IP In Metaverse," Mondaq (Aug. 2, 2022), https://www.mondaq.com/india/trademark/1217914/importance-of-ip-in-metaverse.
95. "Siemens Metaverse Leaks Sensitive Corporate Data: Millions Exposed!," Secure Blink (Apr. 17, 2023), https://www.secureblink.com/cyber-security-news/siemens-metaverse-leaks-sensitive-corporate-data-millions-exposed.
96. Joshi, *supra* note 94.
97. *See* Gayathri Prajit, "Protecting Intellectual Property in the Metaverse: Challenges, Opportunities, and Recent Case Laws," *Times of India* (May 6, 2023), https://toiblogs.indiatimes.com/blogs/voices/protecting-intellectual-property-in-the-metaverse-challenges-opportunities-and-recent-case-laws/.
98. *See* Joshi, *supra* note 94.

99. *See* "4 Health Risks from Using Virtual Reality Headsets," Vest, https://vesttech.com/4-health-risks-from-using-virtual-reality-headsets/ (last visited Nov. 9, 2023).
100. Schneier, *supra* note 5, at 17.
101. *See* Nicole Perlroth, "All 3 Billion Yahoo Accounts Were Affected by 2013 Attack," *New York Times* (Oct. 3, 2017), https://www.nytimes.com/2017/10/03/technology/yahoo-hack-3-billion-users.html.
102. *See, for example*, "FTC Charges Deceptive Privacy Practices in Googles Rollout of Its Buzz Social Network," FTC (Mar. 30, 2011), https://www.ftc.gov/news-events/news/press-releases/2011/03/ftc-charges-deceptive-privacy-practices-googles-rollout-its-buzz-social-network.
103. *See* "Reduce Your Digital Footprint," Carnegie Mellon Information Security Office (Jan. 25, 2021), https://www.cmu.edu/iso/aware/protect-your-privacy/digital-footprint.html.
104. David Huerta & Yael Grauer, "The Best VPN Service," *New York Times* (Mar. 14, 2023), https://www.nytimes.com/wirecutter/reviews/best-vpn-service/.
105. *See* Andy Greenberg, "DuckDuckGo Isn't as Private as You Think," *Wired* (May 28, 2022), https://www.wired.com/story/duckduckgo-microsoft-twitter-ft-bush-assassination-whatsapp/.
106. *See* PwC, *supra* note 52.
107. Ken Hess, "10 Security Best Practice Guidelines for Consumers," *ZDNet* (Mar. 5, 2013), http://www.zdnet.com/10-security-best-practice-guidelines-for-consumers-7000012171/; "Microsoft Protect," Microsoft, http://www.microsoft.com/protect/fraud/phishing/feefraud.aspx (last visited Feb. 1, 2014).

Chapter 5
1. "The Best Metaverse Quotes," Supply Chain Today (last visited Feb. 8, 2024), https://www.supplychaintoday.com/best-metaverse-quotes/.
2. Debra Kamin, "Investors Snap Up Metaverse Real Estate in a Virtual Land Boom," *New York Times* (Dec. 3, 2021), https://www.nytimes.com/2021/11/30/business/metaverse-real-estate.html.
3. More generally, in the context of digital technology, an avatar is "a static or moving image or other graphic representation that acts as a proxy for a person or is associated with a specific digital account or identity, as on the internet." "avatar" Dictionary.com (last visited Feb. 9, 2024).
4. Nintendo of America, "Super Mario Odyssey—Nintendo Switch Presentation 2017 Trailer," YouTube (Jan 13, 2017), https://youtu.be/5kcdRBHM7kM?si=HIurbk2eTfr0Zo7N.
5. HALO, "Halo Infinite | Campaign Launch Trailer," YouTube (Nov 30, 2021), https://youtu.be/PyMlV5_HRWk?si=tluM1Ub05E3KzHV1.
6. Chanel Ferguson, "How Character Creation Works in Final Fantasy XIV," Gamepur (Dec. 7, 2020), https://www.gamepur.com/guides/how-character-creation-works-in-final-fantasy-xiv.
7. "Introducing the Meta Avatars Store," Meta Newsroom (June 20, 2022), https://about.fb.com/news/2022/06/introducing-the-meta-avatars-store/.
8. "Introducing Apple Vision Pro: Apple's first spatial computer," Apple Newsroom (June 5, 2023), https://www.apple.com/newsroom/2023/06/introducing-apple-vision-pro/.

9. Ibid.
10. Ibid.
11. Emma Roth, "Apple's Vision Pro Headset Will Turn You into a Digital Avatar When FaceTiming," *Verge* (June 5, 2023), https://www.theverge.com/2023/6/5/23750096/apple-vision-pro-headset-persona-facetime.
12. "Introducing the Meta Avatars Store," see *supra* note 7.
13. This breakdown of the literature is based on the analysis found in Marko Teräs, "The Lived Experience of Virtual Environments: A Phenomenological Study" (PhD dissertation, Curtin University Business School, 2017), 28. https://espace.curtin.edu.au/handle/20.500.11937/69353.
14. Marie-Laure Ryan, "Narrative as Virtual Reality: Immersion and Interactivity in Literature and Electronic Media," 14 (2003).
15. Bob G. Witmer & Michael J. Singer, "Measuring Presence in Virtual Environments: A Presence Questionnaire," *Presence* 7, no. 3 (1998): 225, 227.
16. Teräs, *supra* note 13, at 31.
17. Ibid. (citing Mel Slater et al., "How We Experience Immersive Virtual Environments: The Concept of Presence and Its Measurement," *Anuario de Psicología* 40 no. 2773 (2009): 193, 205).
18. Janet H. Murray, "Hamlet on the Holodeck," 153 (2000).
19. In first-person shooters, the experience of embodiment already occurs. *See, for example*, Timothy Crick, "The Game Body: Toward a Phenomenology of Contemporary Video Gaming," *Games and Culture* 6, no. 3 (2010): 259, 267 ("I rarely think about controlling the avatar.")
20. Tina L. Taylor, "Living Digitally: Embodiment in Virtual Worlds," in *The Social Life of Avatars: Presence and Interaction in Shared Virtual Environments* ed. Ralph Schroeder (2002), 40, 51.
21. "Express yourself with Meta Avatars," Meta (last visited, Jan 5, 2024), https://www.meta.com/avatars/.
22. Adi Gaskell, "What Our Avatars Say About Us in the Virtual Working World of the Metaverse," *Forbes* (Oct. 11, 2022), https://www.forbes.com/sites/adigaskell/2022/10/11/what-our-avatars-say-about-us-in-the-virtual-working-world-of-the-metaverse/?sh=1931151a4599; summarizing Paul R. Messinger et al., "Reflections of the extended self: Visual self-representation in avatar-mediated environments," *Journal of Business Research* 100 (2019): 531.
23. Do Youn Kim et al., "Avatar-Mediated Experience in the Metaverse: The Impact of Avatar Realism on User-Avatar Relationship," *Journal of Retailing and Consumer Services* 73 (2023): 103382.
24. "Ending Violence and Criminal Sanctions Based on Sexual Orientation and Gender identity: Statement by the High Commissioner," United Nations Human Rights Office (Sep. 17, 2010), https://www.ohchr.org/en/statements/2011/02/ending-violence-and-criminal-sanctions-based-sexual-orientation-and-gender.
25. crash_syndicate, "Non-Human Avatar Options," Meta Community Forums (June 4, 2019), https://communityforums.atmeta.com/t5/eas/Non-Human-Avatar-Options/idi-p/901648.
26. Ibid.
27. Michael Thomsen, "How Does The Sims 4 Handle Gender and Racism?," *Forbes*

(Aug. 15, 2014), https://www.forbes.com/sites/michaelthomsen/2014/08/15/how-does-the-sims-4-handle-gender-and-racism/.
28. "EA's Reckoning: Racism in The Sims 4," Indie Game Atlas (Aug. 17, 2020), https://www.indiegameatlas.com/post/ea-s-reckoning-racism-in-the-sims-4.
29. Nick Yee, "About One Out of Three Men Prefer Playing Female Characters. Rethinking the Importance of Female Protagonists in Video Games," Quantic Foundry (Aug. 5, 2021), https://quanticfoundry.com/2021/08/05/character-gender/.
30. Rosa Mikeal Martey et al., "The Strategic Female: Gender-Switching and Player Behavior in Online Games," Information, Communication, and Society 17, no. 3 (2014): 286, 298.
31. Kimberly M. Christopherson, "The Positive and Negative Implications of Anonymity in Internet Social Interactions: 'On the Internet, Nobody Knows You're a Dog,'" Computers in Human Behavior 23 (2007): 3038, 3040.
32. Ibid.
33. Ibid at 3045.
34. Ibid at 3045-30466.
35. Craig R. Scott, "Benefits and Drawbacks of Anonymous Online Communication: Legal Challenges and Communicative Recommendations," Free Speech Yearbook 41 (2004): 127, 132-133.
36. See, for example, McIntyre v. Ohio Elections Comm'n, 514 U.S. 334, 342 (1995) ("[A]n author's decision to remain anonymous . . . is an aspect of the freedom of speech protected by the First Amendment.").
37. Talley v. California, 362 U.S. 60, 64-65 (1960).
38. Note that this discussion bypasses the current discussion of whether First Amendment protections apply in the context of privately owned online platforms; under current law, they do not.
39. Scott, supra note 35, at 130.
40. Ibid.
41. Ibid., at 131.
42. Ibid.
43. "diversity" Merriam-Webster, https://www.merriam-webster.com/dictionary/diversity (last visited Jan. 7, 2024).
44. James A. Millward, The Silk Road: A Very Short Introduction (2013), 2 (defining the Silk Road as the "ancient transcontinental integration" between "Europe, Southwest Asia (the Middle East), Persia, India, China, and Southeast and Central Asia.")
45. Jan van Dijk, The Digital Divide (Cambridge: Polity Press, 2020), 1 (noting that, despite numerous distinct technical definitions, the "digital divide" generally refers to the division between those who have use, and have access to, information and communications technologies and those who do not.).
46. See, for example, Jessa Lingel, "A Queer and Feminist Defense of Being Anonymous Online," Proceedings of the 54th Hawaii Int'l Conference on System Sciences (2021): 2534, 2534 ("[A]nonymity has inarguable connections to biased and misogynist behavior, it can also be part of a queer and feminist digital toolkit.")

Chapter 6

1. "AI Makes Rules for the Metaverse Even More Important," Brookings (July 13, 2023), https://www.brookings.edu/articles/ai-makes-rules-for-the-metaverse-even-more-important/.
2. "Volume of Data/Information Created, Captured, Copied, and Consumed Worldwide from 2010 to 2020, with Forecasts from 2021 to 2025," Statista, https://www.statista.com/statistics/871513/worldwide-data-created/ (last visited Feb. 9, 2024).
3. "Guidance on the Protection of Personal Identifiable Information," US Department of Labor, https://www.dol.gov/general/ppii (last visited Feb. 19, 2024).
4. "What Personal Data Do Companies Track?," McAfee (Aug. 2, 2022), https://www.mcafee.com/blogs/tips-tricks/what-personal-data-do-companies-track
5. Ibid.
6. Ibid.
7. "What does Privacy Mean?" IAPP, https://iapp.org/about/what-is-privacy (last visited Feb. 19, 2024).
8. Roger Clark, "Introduction to Dataveillance and Information Privacy, and Definitions of Terms," rogerclarke.com, (July 24, 2016), https://www.rogerclarke.com/DV/Intro.html.
9. France Bélanger & Robert E. Crossler, "Privacy in the Digital Age: A Review of Information Privacy Research in Information Systems," *MIS Quarterly* 35 (2011): 1017.
10. "Combining Large Data Sets Challenges IRBs, Researchers to Ensure Privacy," Relias Media (Sep. 1, 2020), https://www.reliasmedia.com/articles/146763-combining-large-data-sets-challenges-irbs-researchers-to-ensure-privacy.
11. "Apple Security Releases," https://support.apple.com/en-us/HT201222 (last visited Oct. 23, 2023).
12. Benjamin Mayo, "What does 'Ask App Not to Track' Mean?," 9TO5Mac (Apr. 27, 2021), https://9to5mac.com/2021/04/27/what-does-ask-app-not-to-track-mean/.
13. Michael Fisher, "The Metaverse and Consumer Data: Here's What you Need to Know," *Drum* (July 21, 2022), https://www.thedrum.com/opinion/2022/07/21/the-metaverse-and-consumer-data-here-s-what-you-need-know.
14. Bailey Martin, "Privacy in a Programmed Platform: How the General Data Protection Regulation Applies to the Metaverse," *Harvard Journal of Law and Technology* 36 no. 235 (2022).
15. "Privacy in the Metaverse," Tsaaro Consulting, https://tsaaro.com/white_paper/privacy-in-the-metaverse/ (last visited Feb. 9, 2024).
16. "Metaverse," European Data Protection Supervisor, https://edps.europa.eu/press-publications/publications/techsonar/metaverse_en (last visited Feb. 9, 2024).
17. Ibid.
18. Ibid.
19. Ibid.
20. Hossein Rahnama & Alex "Sandy" Pentland, "The New Rules of Data Privacy," *Harvard Business Review* (Feb. 25, 2022), https://hbr.org/2022/02/the-new-rules-of-data-privacy.

21. Michael R. Baye & Jeff T. Prince, Managerial Economics and Business Strategy 10th ed., (2022), 3.
22. Blake Morgan, "The 20 Most Compelling Examples of Personalization," *Forbes* (Mar. 29, 2021), https://www.forbes.com/sites/blakemorgan/2021/03/29/the-20-most-compelling-examples-of-personalization/?sh=4957684771b1.
23. Bhugaonkar, Kunal et al., "The Trend of Metaverse and Augmented & Virtual Reality Extending to the Healthcare System," *Cureus* 14 no. 9 (2022).
24. Martin Petkov, "Virtual Health in the Metaverse: Revolutionizing Patient Experiences," Landvault (Jul. 18, 2023), https://landvault.io/blog/metaverse-healthcare.
25. Jeff Prince & Scott Wallsten, "How Much is Privacy Worth Around the World and Across Platforms?," *TPRC48: 48th Research Conference on Communication, Information, and Internet Policy* (Jan. 30, 2020), https://papers.ssrn.com/sol3/papers.cfm?abstract_id=3528386.
26. Ibid.
27. In reality, rather than just determining a proper or improper diagnosis, the different data-sharing scenarios affect the probability of a proper diagnosis. To reflect this reality, one could replace "Proper Diagnosis" with "Proper Diagnosis More Likely" and replace "Improper Diagnosis" with "Proper Diagnosis Less Likely" in Table 6.3.
28. "The Sign of Data Externalities," MD4SG, https://www.md4sg.com/workshop/faact21/the_sign_of_data_externalities.pdf (last visited Feb. 19, 2024).
29. Daron Acemoglu et al., "Too Much Data: Prices and Inefficiencies in Data Markets," *American Economic Journal: Microeconomics* 14 no. 4 (Nov. 2022): 218–256.
30. "BISSELL CleanView Compact Upright Vacuum, Fits In Dorm Rooms & Apartments, Lightweight with Powerful Suction and Removable Extension Wand, 3508, Red, black," https://www.amazon.com/CleanView-Apartments-Lightweight-Removable-3508/dp/B09V5NPHP3/ref=sr_1_3?crid=271SMC9XZAFTO&keywords=vacuum&qid=1698693311&sprefix=vacuum%2Caps%2C127&sr=8-3&th=1 (last visited Feb. 19, 2024).
31. Tim Fisher, "How to Check Real-Time Traffic in Google Maps," *Lifewire* (May 22, 2023), https://www.lifewire.com/check-real-time-traffic-google-maps-7486702.
32. Sophie Putka, "How to Report Traffic Incidents, Hazards, and Speed Checks in Apple Maps on your iPhone," *Business Insider* (Apr. 30, 2021), https://www.businessinsider.com/guides/tech/how-to-report-traffic-incidents-in-maps-on-iphone.
33. Ben Tinker, "How Facebook 'Likes' Predict Race, Religion and Sexual Orientation," *CNN* (Apr 11, 2018), https://www.cnn.com/2018/04/10/health/facebook-likes-psychographics/index.html.
34. "Data Protection and Privacy Legislation Worldwide," UN Trade and Development, https://unctad.org/page/data-protection-and-privacy-legislation-worldwide (last visited Feb. 19, 2024).
35. Prince & Wallsten, *supra* note 25.
36. "Data Protection and Privacy Laws," World Bank, https://id4d.worldbank.org/guide/data-protection-and-privacy-laws (last visited Feb. 9, 2024).
37. Thorin Klosowski, "The State of Consumer Data Privacy Laws in the US (And Why It Matters)," *New York Times* (Sep. 6, 2021), https://www.nytimes.com/wirecutter/blog/state-of-privacy-laws-in-us/.

38. Michael Hill, "The Biggest Data Breach Fines, Penalties, and Settlements So Far," CSO (Sep. 18, 2023), https://www.csoonline.com/article/567531/the-biggest-data-breach-fines-penalties-and-settlements-so-far.html#:~:text=Now%2C%20the%20E quifax%20fine%20has,Regulation%20(GDPR)%20in%20Europe.
39. Klosowski, *supra* note 37.
40. Ibid.
41. Martin, *supra* note 14.
42. Ibid.
43. Ibid.
44. Ibid.
45. "Data Privacy Concerns will be Amplified by the Metaverse," *Verdict* (Jan. 20, 2023), https://www.verdict.co.uk/data-privacy-metaverse-challenge.
46. Paulius Jurcys et al., "Ownership of User-Held Data: Why Property Law is the Right Approach," *JOLT Digest* (Sep. 21, 2021), https://jolt.law.harvard.edu/digest/ownership-of-user-held-data-why-property-law-is-the-right-approach.
47. Rob Frieden, "An Introduction to Data Property Ownership Rights and Data Protection Responsibilities" (Aug. 8, 2019), SSRN, https://papers.ssrn.com/sol3/papers.cfm?abstract_id=3432422.
48. Ibid.
49. Vivien Foster et al., "World Development Report 2021: Data for Better Lives (English)," World Bank Group (June 2021): 224, https://documents1.worldbank.org/curated/en/248201616598597113/pdf/World-Development-Report-2021-Data-for-Better-Lives.pdf.
50. "Data Protection and Privacy Laws," World Bank, https://id4d.worldbank.org/guide/data-protection-and-privacy-laws (last visited Feb. 19, 2024); *see also*, "ID4D Practitioner's Guide (English)," World Bank Group (Oct. 2019): 74, http://documents.worldbank.org/curated/en/248371559325561562/ID4D-Practitioner-s-Guide
51. Prince & Wallsten, *supra* note 25.
52. Ibid.
53. Gary Weingarden, "Metaverse and Privacy," IAPP (Aug. 23, 2022), https://iapp.org/news/a/metaverse-and-privacy-2/.

Chapter 7

1. "The Best Metaverse Quotes," Supply Chain Today, https://www.supplychaintoday.com/best-metaverse-quotes/ (last visited Feb. 8, 2024).
2. "Sci/Tech 'God of the Internet' Is Dead," *BBC News* (Oct. 19, 1998), http://news.bbc.co.uk/2/hi/science/nature/196487.stm.
3. Maggie Harrison, "Interesting Theory: Mark Zuckerberg Is Trying to Become God and Build Heaven," *Futurism* (Oct. 27, 2022), https://futurism.com/theory-mark-zuckerberg-build-heaven.
4. *See, for example*, Lisa Kim, "Facebook Announces New Name: Meta," *Forbes* (Oct. 28, 2021), https://www.forbes.com/sites/lisakim/2021/10/28/facebook-announces-new-name-meta/#.
5. "How Nvidia Dominated AI—and Plans to Keep it That Way As Generative AI Explodes," *Venture Beat* (Feb. 23, 2023), https://venturebeat.com/ai/how-nvidia-dominated-ai-and-plans-to-keep-it-that-way-as-generative-ai-explodes/.

6. *See* Steve Lohr, "The Power of the Platform at Apple," *New York Times* (Jan. 29, 2011), https://www.nytimes.com/2011/01/30/business/30unbox.html.
7. *See* "Is Apple Coming up with its own AI Platform?," *Telangana Today* (May 22, 2023), https://telanganatoday.com/is-apple-coming-up-with-its-own-ai-platform.
8. *See, for example*, "What Laws Govern The Metaverse?," Blockchain Council (Jan. 16, 2023), https://www.blockchain-council.org/metaverse/what-laws-govern-the-metaverse/#:~:text=To%20make%20things%20somewhat%20better%2C%20the%20Metaverse%20realm%20has%20accepted%20to%20abide%20by%20the%20general%20laws%20that%20apply%20to%20the%20web%20so%20as%20to%20bring%20a%20sense%20of%20uniformity%2C%20security%2C%20and%20transparency%20within%20its%20ecosystem.
9. *See* Tom Ara et al., "Exploring the Metaverse: What Laws will Apply?," DLA Piper (Feb. 22, 2022), https://www.dlapiper.com/en/insights/publications/2022/02/exploring-the-metaverse.
10. Ibid.
11. Ibid.
12. Ibid.
13. Blockchain Council, *supra* note 8.
14. Ibid.; Niveditha Jain, "Market Review: The Rise and Fall of NFTs in 2022," *Auction Daily* (Dec. 22, 2022), https://auctiondaily.com/news/market-review-the-rise-and-fall-of-nfts-in-2022/.
15. "What Role does Spatial Computing Play in the Metaverse?," IEEE Metaverse, https://metaversereality.ieee.org/publications/articles/the-role-spacial-computing-plays-in-the-metaverse#:~:text=Spatial%20computing%20significantly%20affects%20the,physical%20elements%20coexist%20and%20interact (last visited July 9, 2024).
16. *See* Timothy Marler et al., "The Metaverse and Homeland Security Opportunities and Risks of Persistent Virtual Environments," RAND (May 2023), https://www.rand.org/pubs/perspectives/PEA2217-2.html.
17. For a deeper dive, *see* Scott J. Shackelford & Scott O. Bradner, *Forks in the Digital Road: Key Decisions that Gave Us the Internet we Have* (2024).
18. *See, for example*, John Perry Barlow, "A Declaration of Cyberspace Independence" (Feb. 8, 1996), https://projects.eff.org/~barlow/Declaration-Final.html. For a good overview, *see* Jack Goldsmith & Tim Wu, *Who Controls the Internet? Illusions of a Borderless World* (2008).
19. *See* Bethany Allen-Ebrahimian, "The Man Who Nailed Jello to the Wall," *Foreign Policy* (June 29, 2016), https://foreignpolicy.com/2016/06/29/the-man-who-nailed-jello-to-the-wall-lu-wei-china-internet-czar-learns-how-to-tame-the-web/.
20. Ibid.
21. Scott J Shackelford & Amanda N. Craig, "Beyond the New Digital Divide: Analyzing the Evolving Role of National Governments in Internet Governance and Enhancing Cybersecurity Symposium," *Stanford Journal of International Law* 50 (2014): 119.
22. Jonathan Masters, "What Is Internet Governance?," CFR (Apr. 23, 2014), https://www.cfr.org/backgrounder/what-internet-governance.

23. Larry Downes, "Requiem for Failed UN Telecom Treaty: No One Mourns the WCIT," *Forbes* (Dec. 17, 2012), http://www.forbes.com/sites/larrydownes/2012/12/17/no-one-mourns-the-wcit/.
24. "About," Web 3 Foundation, https://web3.foundation/about/ (last visited June 23, 2022).
25. *See* Tom Barrett, "Will Web3 Make ICANN Obsolete?," *Forbes* (July 13, 2022), https://www.forbes.com/sites/forbesbusinesscouncil/2022/07/13/will-web3-make-icann-obsolete/?sh=77445c51d286.
26. João Marinotti, "Can You Truly Own Anything in the Metaverse? A Law Professor Explains How Blockchains and NFTs Don't Protect Virtual Property," Conversation (Apr. 21, 2022), https://theconversation.com/can-you-truly-own-anything-in-the-metaverse-a-law-professor-explains-how-blockchains-and-nfts-dont-protect-virtual-property-179067.
27. Ibid.
28. Ara et al., *supra* note 9.
29. Ibid.
30. Ibid.
31. Ibid.
32. *See* Cathy Li, "Who will Govern the Metaverse?," World Economic Forum (May 25, 2022), https://www.weforum.org/agenda/2022/05/metaverse-governance/.
33. Ibid.
34. Sam Jungyun Choi et al., "Regulating the Metaverse in Europe," Inside Privacy (Apr. 12, 2023), https://www.insideprivacy.com/eu-data-protection/regulating-the-metaverse-in-europe/#:%7E:text=A%20key%20aspect%20of%20the,certain%20types%20of%20crypto%2Dassets.
35. Ibid.
36. *See* Li, *supra* note 30.
37. Ibid.
38. Aviv Ovadya, "Meta Ran a Giant Experiment in Governance. Now It's Turning to AI," *Wired* (July 10, 2023), https://www.wired.com/story/meta-ran-a-giant-experiment-in-governance-now-its-turning-to-ai/.
39. Ibid.
40. Ibid.
41. An earlier version of this section was published as Scott J. Shackelford, "Companies' Self-Regulation Doesn't Have to Be Bad for the Public," Conversation (June 12, 2019), https://theconversation.com/companies-self-regulation-doesnt-have-to-be-bad-for-the-public-117565.
42. *See, for example*, Kenneth J. Arrow et al., "Elinor Ostrom: An Uncommon Woman for the Commons," *PNAS* 109 (2012): 13135–13136, https://www.pnas.org/doi/10.1073/pnas.1210827109.
43. Garett Hardin, "The Tragedy of the Commons," *Science* 162 no. 1243, (1968): 1244–1245.
44. Michael D. McGinnis, "An Introduction to IAD and the Language of the Ostrom Workshop: A Simple Guide to a Complex Framework," *Policy Studies Journal* 39 no. 163 (2011): 171–72, http://php.indiana.edu/~mcginnis/iad_guide.pdf.

45. Michael D. McGinnis, "Costs and Challenges of Polycentric Governance: An Equilibrium Concept and Examples from U.S. Health Care," Workshop on Self-Governance, Polycentricity, and Development in *Conference on Self-Governance, Polycentricity, and Development* (Beijing: Renmin University, China 2011), 1–2, http://php.indiana.edu/~mcginnis/Beijing_core.pdf.
46. *See* Elinor Ostrom, "Polycentric Systems as One Approach for Solving Collective-Action Problems" *Indiana University Workshop in Political Theory and Policy Analysis*, (Working Paper Series No. 08–6, 2008): 1–2, http://dlc.dlib.indiana.edu/dlc/bitstream/handle/10535/4417/W08-6_Ostrom_DLC.pdf.
47. *See* David Feeny et al., "The Tragedy of the Commons: Twenty-Two Years Later," *Human Ecology* 1, no. 4 (1990): 18 (describing the open access system of property rights as one in which access to the resource is available to everyone, free, and unregulated). Feeny also explains that open access systems lead to degradation of the resource due to overuse and an inability to enforce regulations or exclusion mechanisms. Ibid., at 6, 9.
48. *See* Sarah Knapton, "International Space Station Astronauts Plug Leak with Finger and Tape After Being Hit with Debris," *Telegraph* (Aug. 31, 2018), https://www.telegraph.co.uk/science/2018/08/30/international-space-station-leaking-air-hit-space-debris/. However, it later emerged that a drill may have been to blame. *See* Thomas McMullan, "Was the Hole on the International Space Station Really Sabotage?," *Wired* (Sep. 13, 2018), https://www.wired.co.uk/article/international-space-station-leak-hole-drill-what-happened; "ISS: Spacewalk Cosmonauts Investigate Mystery Hole," *BBC* (Dec. 11, 2018), https://www.bbc.com/news/science-environment-46529422.
49. Thierry Sénéchal, "Orbital Debris: Drafting, Negotiating, Implementing a Convention" (unpublished M.B.A. thesis, Sloan School of Management, 2007), 99, http://hdl.handle.net/1721.1/39519.
50. "Recommendations for Regulating AI," Google, https://ai.google/static/documents/recommendations-for-regulating-ai.pdf.
51. *See* Kaitlyn Washburn, "Nearly 43,000 People Died from Gun Violence in 2023: How to Tell the Story," Association of Healthcare Journalists (Feb. 14, 2024), https://healthjournalism.org/blog/2024/02/nearly-43000-people-died-from-gun-violence-in-2023-how-to-tell-the-story/.
52. *See, for example*, "Inflation, Health Costs, Partisan Cooperation Among the Nation's Top Problems," Pew Research Center (June 21, 2023), https://www.pewresearch.org/politics/2023/06/21/inflation-health-costs-partisan-cooperation-among-the-nations-top-problems/.
53. *See* Robert Caserta, "The Metaverse: Why Governments Should Care," KPMG (Jan. 18, 2023), https://kpmg.com/xx/en/blogs/home/posts/2023/01/the-metaverse-why-governments-should-care.html.
54. Brian X. Chen, "What's All the Hype About the Metaverse?," *New York Times* (Jan. 18, 2022), https://www.nytimes.com/2022/01/18/technology/personaltech/metaverse-gaming-definition.html.
55. "The Metaverse: Concepts and Issues for Congress," CRS R47224 (Aug. 26, 2022), 18–23, https://crsreports.congress.gov/product/pdf/R/R47224; FCC, "Fourteenth Broadband Deployment Report," FCC 21-18 (Jan. 19, 2021), 19–20, https://docs.fcc.gov/public/attachments/FCC-21-18A1.pdf.

56. "The Metaverse: Concepts and Issues for Congress," *supra* note 55, at 1 (citing Information Infrastructure and Technology Act of 1992 (S. 2937 and H.R. 5759, the 102 Congress)).
57. For an introduction to blockchain governance challenges, *see* Scott J. Shackelford & Steve Myers, "Block-by-Block: Leveraging the Power of Blockchain Technology to Build Trust and Promote Cyber Peace," *Yale Journal of Law and Technology* 19 (2017): 334.
58. "The Metaverse: Concepts and Issues for Congress," *supra* note 55, at 3.
59. *See, for example*, Hannah Murphy, "How Will Facebook Keep Its Metaverse Safe for Users?," *Financial Times* (Nov. 12, 2021), https://www.ft.com/content/d72145b7-5e44-446a-819c-51d67c5471cf.
60. Tom Wheeler, "The Metachallenges of the Metaverse," Brookings (Sep. 30, 2021), https://www.brookings.edu/blog/techtank/2021/09/30/the-metachallenges-of-the-metaverse/.
61. Andrew Donnelly, "Reps. Blunt Rochester, Walberg Introduce Bipartisan Legislation to Build the Workforce of the Future," US House of Representatives press release (May 11, 2023), https://bluntrochester.house.gov/news/documentsingle.aspx?DocumentID=2982#:~:text=The%20Immersive%20Technology%20for%20the%20American%20Workforce%20Act%20would%3A,workforce%20development%20utilizing%20immersive%20technology; Other proposed bills include those related to innovation, competition, kids safety, and digital oversight that would empower the FTC to take a larger role in policing the Metaverse; "The Metaverse: Concepts and Issues for Congress," *supra* note 55, at 25.
62. "The Metaverse: Concepts and Issues for Congress," *supra* note 55, at 6.
63. Ibid., at 7.
64. "The Metaverse," *supra* note 55.
65. White House, "A Declaration for the Future of the Internet," (Apr. 28, 2022), at 1, https://www.whitehouse.gov/wp-content/uploads/2022/04/Declaration-for-the-Future-for-the-Internet_Launch-Event-Signing-Version_FINAL.pdf.
66. Ibid., at 12.

Chapter 8

1. Diana Ambolis, "10 Best Metaverse Quotes Everyone Should Read," *Blockchain Magazine* (May 25, 2023), https://blockchainmagazine.net/10-best-metaverse-quotes-everyone-should-read/#:~:text=The%20metaverse%20can%20be%20summed,been%20defined%20by%20several%20people.
2. *See* Nicholas Crafts, "Artificial Intelligence as a General-Purpose technology: An Historical Perspective," Oxford Review of Economic Policy 37, no. 3 (2021): 521; Bernard Marr, "What Is the Artificial Intelligence Revolution and Why Does It Matter to Your Business?," *Forbes* (Aug.10, 2022), https://www.forbes.com/sites/bernardmarr/2020/08/10/what-is-the-artificial-intelligence-revolution-and-why-does-it-matter-to-your-business/?sh=708d31cd547b.
3. Doug Austin, "The Metaverse is Dead, Long Live Generative AI," eDiscoveryToday (May 10, 2023) https://ediscoverytoday.com/2023/05/10/the-metaverse-is-dead-long-live-generative-ai-artificial-intelligence-trends/ agreeing with Ed Zitron, "RIP Metaverse," *Business Insider* (May 8, 2023), https://www.businessinsider.com/metaverse-dead-obituary-facebook-mark-zuckerberg-tech-fad-ai-chatgpt-2023-5.

4. Tim Baysinger, "Metaverse Funding Plummets as Investors Favor Generative AI," *Axios* (Mar. 16, 2023), https://www.axios.com/pro/media-deals/2023/03/16/metaverse-funding-plummets-as-investors-favor-generative-ai.
5. Mathieu Nouzareth, "How Generative AI Can Play a Role in the Metaverse," *Nasdaq* (Apr. 11, 2023), https://www.nasdaq.com/articles/how-generative-ai-can-play-a-role-in-the-metaverse.
6. Ibid.
7. Ibid.
8. Tom Wheeler, "AI Makes Rules for the Metaverse Even More Important," *Brookings* (July 13, 2023), https://www.brookings.edu/articles/ai-makes-rules-for-the-metaverse-even-more-important/.
9. Tomas Chamorro-Premuzic, "How Different Are Your Online and Offline Personalities?," *Guardian* (Sep. 24, 2015), https://www.theguardian.com/media-network/2015/sep/24/online-offline-personality-digital-identity.
10. John Suler, "The Dimensions of Cyberpsychology Architecture," in *Boundaries of Self and Reality Online*, eds. Jayne Gackenbach & Johnathan Bown (London: Academic, 2017), 1–23.
11. *See* Chamorro-Premuzic, *supra* note 9.
12. *See* Wheeler, *supra* note 7.
13. Natalie Neysa Alund, "Why do People Catfish? What are the Signs of It? Here's What You—and Your Kids—Should Know," *USA Today* (Nov. 29, 2022), https://www.usatoday.com/story/news/nation/2022/11/29/what-is-catfishing-why-do-people-catfish/10794619002/.
14. "How to Recognize and Avoid Phishing Scams," Federal Trade Commission (Sep. 2022), https://consumer.ftc.gov/articles/how-recognize-and-avoid-phishing-scams.
15. Randy Ginsburg, "AI and the Metaverse: 6 Ways AI Will Unlock the Metaverse," Touchcast (June 29, 2022), https://touchcast.com/blog-posts/ai-and-the-metaverse.
16. Ibid. *See also*,Carolyn E. Pepper & Jonathan J. Andrews, "Reed Smith Guide to the Metaverse," Reed Smith LLP 1, (2022), 98–99 (defining deepfakes and shallowfakes and the concerns raised by their use in the Metaverse).
17. Wu Haojie et al., "Deepfake in the Metaverse: An Outlook Survey" 1 (June 12, 2023), https://arxiv.org/pdf/2306.07011.pdf (discussing the security threats wrought by deepfakes in the Metaverse).
18. Ginsburg, *supra* note 15.
19. Wu et al., *supra* note 17.
20. Elizabeth de Luna, "Zuckerberg Backtracks after Horizon Worlds Backlash, Claims Meta Is 'Capable of Much More'," *Mashable* (Aug. 19, 2022), https://mashable.com/article/mark-zuckerberg-horizon-worlds-backlash-meta.
21. Ibid.
22. Dani Di Placido, "Mark Zuckerberg Upgraded His 'Metaverse' Avatar after the Entire Internet Laughed at Him," *Forbes* (Aug. 21, 2022), https://www.forbes.com/sites/danidiplacido/2022/08/21/mark-zuckerberg-upgraded-his-metaverse-avatar-after-the-entire-internet-laughed-at-him/.
23. Nicole Clark, "Mark Zuckerberg Updates his Metaverse Avatar to Look Slightly more Human," *Polygon* (Aug. 19, 2022), https://www.polygon.com/23313564/mark-zuckerberg-metaverse-horizon-worlds-graphics-update.
24. Ma Shugao et al., "Pixel Codec Avatars," Meta (June 19, 2021), https://research.facebook.com/publications/pixel-codec-avatars/.

25. Image found on Ben Wodecki, "Meta Connect 2022: Photorealistic AI-powered Avatars," AI Business (Oct. 11, 2022), https://aibusiness.com/verticals/meta-connect-2022-photorealistic-ai-powered-avatars.
26. Matt Growcoot, "Meta's Photorealistic Avatars Can be Generated with Just an iPhone," PetaPixel (Jun 14, 2022) https://petapixel.com/2022/06/14/metas-photorealistic-avatars-can-be-generated-with-just-an-iphone/.
27. Ibid.
28. José Adorno, "Here's how Apple Vision Pro Digital Personal Will Work," *BGR* (Oct. 6, 2023), https://bgr.com/tech/heres-how-apple-vision-pro-digital-persona-will-work/.
29. Malcolm Owen, "Apple isn't Standing Still on Generative AI, and Making Human Models Dance is Proof," Apple Insider (Dec. 19, 2023), https://appleinsider.com/articles/23/12/19/apple-isnt-standing-still-on-generative-ai-and-making-human-models-dance-is-proof.
30. Ibid.
31. Corinna Lathan, "Generative AI in the Metaverse," *Psychology Today* (June 21, 2023), https://www.psychologytoday.com/us/blog/inventing-the-future/202306/generative-ai-in-the-metaverse.
32. Victor Erukhimov, "The Path to Photorealistic Avatars," *Forbes* (Dec. 11, 2023), https://www.forbes.com/sites/forbestechcouncil/2023/12/11/the-path-to-photorealistic-avatars.
33. *See, generally*, Christopher Cozzens, "The Patchwork Privacy Problem: How the United States' Privacy Regime Fails to Protect its Businesses and Data Subjects," *Seton Hall Law Review* 52 (2021): 1157.
34. Anders Christoffersen et al., "Level Up: The Future of Video Games Is Bright," Bain & Co. (Oct. 12, 2022), https://www.bain.com/insights/level-up-the-future-of-video-games-is-bright/.
35. "AAA Cost of Development: How Expensive Are these Games?," Main Leaf (June 20, 2023), https://mainleaf.com/aaa-cost-of-development-how-expensive-are-they/.
36. Ibid.
37. Frederick Daso, "Kaedim Helps Game Developers Generate 3D Models Using AI," *Forbes* (Aug. 9, 2022), https://www.forbes.com/sites/frederickdaso/2022/08/09/yc-backed-kaedim-helps-game-developers-generate-3d-models-using-ai/?sh=780d06bd646e.
38. Ibid.
39. Thomas Wilde, "How Generative AI Could Change the Way Video Games are Developed, Tested, and Played," Geek Wire (Dec. 27, 2023), https://www.geekwire.com/2023/how-generative-ai-could-change-the-way-video-games-are-developed-tested-and-played/.
40. "This Is How GenAI Could Accelerate the Metaverse," Boston Consulting Group (Aug. 16, 2023), https://www.bcg.com/publications/2023/how-gen-ai-could-accelerate-metaverse.
41. Dean Takahashi, "How Generative AI Could Create Assets for the Metaverse," *VentureBeat: GamesBeat* (Nov. 28, 2022), https://venturebeat.com/games/how-generative-ai-could-create-assets-for-the-metaverse-jensen-huang/.
42. Will Knight, "Why It's So Hard to Count Twitter Bots," *Wired* (May 18, 2022), https://www.wired.com/story/twitter-musk-bots/.
43. Ibid.

44. "Twitter Bot: Get Auto Followers, Auto Retweet & Likes," Tweetfull (July 21, 2020), https://tweetfull.com/blog/twitter-bot-how-to-make/.
45. "How to Combat Twitter Bots and Twitter Misinformation," Peakmetrics, https://www.peakmetrics.com/insights/how-to-combat-twitter-bots-and-twitter-misinformation (last visited Jan. 1, 2024).
46. Ibid.
47. "Tweet Smarter, Not Harder: Using ChatGPT to Automate Twitter Conversations," Ghost Retail (Sep. 7, 2023), https://www.ghostretail.com/post/using-chatgpt-to-automate-twitter-conversations.
48. Ibid.
49. Philipp Pointner, "Generative AI: Risks and Solutions for a Safer Digital Landscape," Tech Radar (Aug. 1, 2023), https://www.techradar.com/pro/generative-ai-risks-and-solutions-for-a-safer-digital-landscape.
50. "Social Media Bots Overview," National Protection and Programs Directorate Office of Cyber and Infrastructure Analysis (May 2018) https://niccs.cisa.gov/sites/default/files/documents/pdf/ncsam_socialmediabotsoverview_508.pd.
51. Ibid.
52. Ibid.
53. Carlos Melendez, "The Metaverse: Driven by AI, Along with the Old Fashioned Kind of Intelligence," *Forbes* (Apr. 18, 2022), https://www.forbes.com/sites/forbestechcouncil/2022/04/18/the-metaverse-driven-by-ai-along-with-the-old-fashioned-kind-of-intelligence/?sh=ff5f1931b365.
54. Boston Consulting Group, see *supra* note 40.
55. Carolyn E. Pepper & Jonathan J. Andrews, "Deepfakes in the Metaverse," Reed Smith (Aug. 1, 2022), https://www.reedsmith.com/en/perspectives/metaverse/2022/08/deepfakes-in-the-metaverse.
56. Seo-Young Chu, "Robot Rights," in *Do Metaphors Dream of Literal Sleep?* (Cambridge: Harvard University Press, 2010), 215 (noting the 1921 Karel Čapek play R.U.R., which explores the idea of the "exploitation of robot slaves.").
57. *See, generally*, "International Copyright Issues and Artificial Intelligence," Copyright.gov (July 26, 2023), https://www.copyright.gov/events/international-ai-copyright-webinar/; John Villasenor, "Patents and AI Inventions: Recent Court Rulings and Broader Policy Questions," Brookings (Aug. 25, 2022), https://www.brookings.edu/articles/patents-and-ai-inventions-recent-court-rulings-and-broader-policy-questions/.
58. Jane Bambauer et al., "Platforms: The First Amendment Misfits," *Indiana Law Journal* 97 no. 3 (2022): 1047 (noting that, currently, "First Amendment precedents that allowed government to require a private entity to host the speech of others have limited applicability to online platforms like Twitter and Facebook.").
59. Toni M. Massaro & Helen Norton, "Siri-ously? Free Speech Rights and Artificial Intelligence," Northwestern University Law Review 110 no. 5 (2016): 1169, 1175–1178.
60. Sarah Cook, "China's Censors could Shape the Future of AI-Generated Content," *Japan Times* (Feb. 27, 2023), https://www.japantimes.co.jp/opinion/2023/02/27/commentary/world-commentary/china-artificial-intelligence/.
61. Norman Eisen et al., "AI Can Strengthen U.S. Democracy—and Weaken It," Brookings (Nov. 21, 2023), https://www.brookings.edu/articles/ai-can-strengthen-u-s-democracy-and-weaken-it/.
62. Ibid.

Chapter 9

1. My Geek Wisdom, https://mygeekwisdom.com/2022/07/30/its-not-safe-out-here-its-wondrous-with-treasures-to-satiate-desires-both-subtle-and-gross-but-its-not-for-the-timid/ (last visited Nov. 11, 2023).
2. *See* Lauren Feiner, "Meta Sued by 33 state AGs for Addictive Features Targeting Kids," *NBC News* (Oct. 24, 2023), https://www.nbcnews.com/tech/tech-news/meta-sued-33-state-ags-addictive-features-targeting-kids-rcna121927#.
3. Ibid.
4. Eric Reicin, "Unfamiliar and Unregulated Territory: Protecting Kids in the Metaverse," *Forbes* (Oct. 4, 2023), https://www.forbes.com/sites/forbesnonprofitcouncil/2023/10/04/unfamiliar-and-unregulated-territory-protecting-kids-in-the-metaverse/?sh=268ce36a4890.
5. Jada Jones, "Meta Lowered the Age Limit for Quest Accounts. Are these kids too young for exploring VR?," *ZDNet* (June 20, 2023), https://www.zdnet.com/article/meta-lowered-the-age-limit-for-quest-accounts-are-these-kids-too-young-for-exploring-vr/.
6. *See* "Parents' Guide to the Metaverse," Security.org, https://www.security.org/digital-safety/parents-metaverse-guide/ (last visited Nov. 11, 2023).
7. *See* Sheera Frenkel & Kellen Browning, "The Metaverse's Dark Side: Here Come Harassment and Assaults," *New York Times* (Dec. 30, 2021), https://www.nytimes.com/2021/12/30/technology/metaverse-harassment-assaults.html.
8. *See* Tayna Basu, "The Metaverse Has a Groping Problem Already," *MIT Technology Review* (Dec. 16, 2021), https://www.technologyreview.com/2021/12/16/1042516/the-metaverse-has-a-groping-problem/.
9. *See* Michal Gromek, "Are We Ready For Avatars Reporting Sexual Harassment In The Metaverse Police Stations?," *Forbes* (May 8, 2023), https://www.forbes.com/sites/digital-assets/2023/05/08/are-we-ready-for-avatars-reporting-sexual-harassment-in-the-metaverse-police-stations/?sh=6077a0de5139.
10. Brittan Heller, "We Need a 911 for the Metaverse," Stanford PACS (Mar. 30, 2022), https://pacscenter.stanford.edu/wp-content/uploads/2022/08/We-Need-a-911-for-the-Metaverse-—-The-Information.pdf.
11. Janna Anderson & Lee Rainie, "Visions of the Internet in 2035," Pew Research Center (Feb. 7, 2022), https://www.pewresearch.org/internet/2022/02/07/visions-of-the-internet-in-2035/.
12. Ibid.
13. Ibid.
14. *See* "Screen Time vs. Lean Time Infographic," CDC, https://www.cdc.gov/nccdphp/dnpao/multimedia/infographics/getmoving.html (last visited Nov. 11, 2023).
15. *See* Elinor Ostrom, *Governing the Commons: The Evolution of Institutions for Collective Action*, 1st ed. (New York: Cambridge University Press, 1990), 211–212; Susan J. Buck, *The Global Commons: An Introduction*, 2nd ed. (Abingdon: Earthscan, 1998), 32.
16. "Interview with Nobel Laureate Elinor Ostrom," Escotet Foundation, https://escotet.org/2010/11/interview-with-nobel-laureate-elinor-ostrom/ (last visited Nov. 11, 2023).

INDEX

For the benefit of digital users, indexed terms that span two pages (e.g., 52–53) may, on occasion, appear on only one of those pages.

abuse, harassment, and bullying, 32–33, 35–36, 44–45, 59–61, 76, 139
access controls, 57–58
accessibility, 77–79
ads, targeted, 94
advertising, 92–93, 138–139
agency, and interactivity, 70
AI. *See* artificial intelligence (AI)
anonymity, 32, 73, 79, 91
antitrust regulations, 24–25
app permissions, 57–58
App Tracking Transparency (ATT) initiative, 88
Apple, 9, 15, 21, 68, 126–127
AR (augmented reality) headsets. *See* headsets
artificial intelligence (AI), 119
   and avatars, 124–127
   bots, 123, 129–135
   and cybersecurity, 52
   environment and asset rendering, 127–129
   and moderation, 40
   and reality, 121–129
   relation to Metaverse, 119–120

asset creation, 127–128
ATT (App Tracking Transparency) initiative, 88
augmented reality (AR) headsets. *See* headsets
avatars. *See also* identity
   and AI, 123–127, 129–135
   and child protection, 45–46
   possibilities and risks, 34–36
   and privacy, 91
   trademark concerns, 42

barriers to entry, 22–26
behavioral data, 82, 94
biases, discrimination, and stereotypes, 73–74
Biasini, Nick, 61–62
BID (biometrically inferred data), 94
biometric information, 54–55, 58, 89–90, 94, 127
blockchains, 52
bots (AI), 123, 129–135
bullying. *See* abuse, harassment, and bullying
"Burning Chrome" (Gibson), 1–2

Index

Carmille, René, 49
catfishing, 36, 122
child protection, 45–46, 60–61, 99, 138–140
Chimeria, 35
China, 1–2, 106–107
CISA (Cybersecurity and Infrastructure Security Agency), 131
Clark, David, 6
Cline, Ernest, 55
Clinton, Bill, 106–107
cloud services, and zero trust security, 51
companies. See firms
complements, 27
compute, 15
Congressional Caucus on Virtual, Augmented, and Mixed Reality (MR) Technologies, 115–116
consumer devices, 15
consumer protection law, 45
content moderation, 38–41, 116–117
Cook, Tim, 4–5
copyright and trademark concerns, 42–43, 61–62
COVID-19 pandemic, 54–55
criminal law and tort law, 44–45
cross-border data transfers, 102
cultural norms, 38–40, 44, 103
cyberbullying. See abuse, harassment, and bullying
cybersecurity, 48
  costs, 48
  cybercrimes, 59–61
  data and privacy protection practices, 58–59
  digital footprints, 53–55, 62–65
  encryption and VPNs, 59
  and interoperability, 20
  learning from failures, 49–52
  strong security practices, 55–58
  unique cyber threats, 61–62

Cybersecurity and Infrastructure Security Agency (CISA), 131
cyberspace, 1–3

DAOs (decentralized autonomous organizations), 110–111
data breaches, 85
data collection, 81–83, 89–90, 92–97. See also privacy
data localization laws, 102–103, 117
data ownership rights, 101–102
data privacy. See privacy
data profiles, 53–55, 62–65
data sharing, 96–98, 101–102
data synchronization challenges, 20
deceit, 122–124
decentralization, 26, 40–41, 107–108, 110–111
decentralized autonomous organizations (DAOs), 110–111
deepfakes, 52, 123–124, 132, 135
defamation law, 46–47
demand, 14, 17–18
development costs, for virtual-world games, 128
digital assets, ownership of, 108
Digital Compass, 117
digital divide, 78–79, 106–108, 115–116
digital footprints, 53–55, 62–65
Digital Services Act and Code for Disinformation, 109
digital sovereignty, 117
direct network effects, 17–18, 21–22
discrimination, biases, and stereotypes, 73–74
disinformation, and cybersecurity, 52
diversity, 72–74, 77–79
Durmus, Murat, 6–7

e-commerce, 16
economics, 10
  coexistence with physical world, 26–30
  one versus many Metaverses, 19–26

economics (*Continued*)
  predictions, 5–9
  scarcity, 10–14
  supply and demand, 14–18
  economies of scale, 20
education, virtual environments for, 34, 37
embodied Internet, 115–116
embodiment, 32, 71, 72, 133
empathy, 33–34, 73
encryption, 59
Engressia, Joe, 49
Entrup, Richard, 5
Errazuriz, Sebastian, 37
European Union (EU), 99, 101, 109, 117
externalities, 96–98, 101–102

Facebook. *See* Meta
fairness, 77–79
Federal Trade Commission (FTC), 51, 53–55, 62
Financial Times calculator, 54
firms, 80–81, 87–88, 91–94, 101–102, 108–111
First Amendment, 43, 76, 133
fixedness, 42
fraudulent behavior, 56–58, 122–124
free speech, 43, 132–135
freedom of expression, 72–75

games and gaming, 16–17, 127–129
GDPR (General Data Protection Regulation), 99, 101
gender, and avatar choice, 74
General Data Protection Regulation (GDPR), 99, 101
generative AI. *See* artificial intelligence (AI)
Gibson, William, 1–2, 141–142
governance, 104
  companies' existing governance, 108–111
  and content moderation, 40
  government regulations, 114–118
  Internet governance, 106–108
  laws governing Metaverse, 105–106 *See also* laws and legal frameworks
  lessons from other contexts, 111–114
  government regulations. *See* laws and legal frameworks

hacking, 49–50. *See also* cybersecurity
haptic technology, and moderation, 39
harassment. *See* abuse, harassment, and bullying
Hardin, Garrett, 112
hardware/software complementarity, 16
Harrell, D. Fox, 35
hate speech, 59–61
headsets, 9, 13–14, 21, 58, 126–127
Huang, Jensen, 128–129

ICANN (Internet Corporation for Assigned Names and Numbers), 107–108
identity, 66
  anonymity, 73–77, 79
  "being" in the Metaverse, 66–69
  diversity, 72–74, 77–79
  embodiment, 71–72
  immersion, 69–70
  interactivity, 70
  personal identity, 71–74
identity theft, 54–55
immersion, 3, 28–30, 32–38, 69–70
Immersive Technology for the American Workforce Act, 116–117
indirect network effects, 17–18, 21–22
information privacy, 83–88
infrastructure, 14–17, 50
intellectual property (IP), 61–62, 105
interactivity, 70
international data sharing, 102
Internet Corporation for Assigned Names and Numbers (ICANN), 107–108
Internet governance, 106–108
interoperability, 19–20, 22–23
IP (intellectual property), 61–62, 105
Isaacson, Walter, 49

jurisdiction, and moderation, 40–41
Koons, Jeff, 37
Krugman, Paul, 4–5

laws and legal frameworks, 23–25, 40–47, 99–103, 105–106, 114–118, 132–135
limit pricing, 24
location privacy preferences, 95–96

McGinnis, Michael, 112
medical treatment, remote, 94–95
Meta, 5, 15, 51, 67, 111
Meta Quest, 21
MetaGuard, 90–91
the Metaverse
  artificial intelligence (AI). *See* artificial intelligence (AI)
  bubble, 7–9, 115–116
  defined, 1–3
  economics. *See* economics
  enthusiasts and pessimists, 5–7
  future of, 136
  governance. *See* governance
  identity. *See* identity
  infrastructure, 13–17
  investment in and usage of, 3–5
  and privacy. *See* privacy
  securing. *See* cybersecurity
  speech and expression in. *See* speech and expression
metaverses
  defined, 2–3
  early platforms, 23
  emergence and barriers to entry, 22–25
  potential forms of governance, 25–26
  in stories and films, 62, 136–138
Metaverse-specific experiences, 16–17
Millward, James A., 78
moderation, 38–41, 116–117
Morris, Robert Tappan, 49–50
multi-factor authentication (MFA), 56–57
multihoming, 20–21

national security, 106
network capacity, 15
network effects, 17–18, 21–22
Neuromancer (Gibson), 2
NFTs (non-fungible tokens), 13, 52, 61–62, 110–111
nonverbal cues and facial expressions, 124–127
North American Electric Reliability Council (NERC), 117–118

online privacy. *See* privacy
Ostrom, Elinor (Lin), 111–113, 118, 142
Ostrom, Vincent, 111–112
overlords, 23–25

passwords, 51, 57, 62
perceptual presence, 69–70
persistence, 3
personal boundaries, 33
personal expression. *See* speech and expression
personal identifiable information (PII), 82, 84–85
personal identity, 71–74
personas, 68, 126–127. *See also* avatars
phishing, 56–58, 122–123
photorealistic content, AI generation of, 128–129
physical world, coexistence with, 26–30
PII (personal identifiable information), 82, 84–85
political expression, and AI bots, 133–135
polycentric governance, 112–114, 117
pop culture, 136–138
price, and scarcity, 11
prisoner's dilemma, 96
privacy, 80
  and anonymity, 75
  compared to other platforms, 88–92
  definition, 81–88
  desirability of, 92–97

privacy (*Continued*)
   and digital footprints, 53–55, 62–65
   and headsets, 58
   possibility of, 97–98
   privacy protection practices, 58–59
   regulation, 99–103
Proteus Effect, 34–35
pseudonymity, 36
psychological manipulation, 33

Quest VR headset, 21

Ready Player One (Cline), 55, 141–142
realism, 36–37, 124–127
reality, 121–129
Reality Caucus, 115–116
recommendations, 93–94, 97–98
regulations. *See* governance; laws and legal frameworks
Roblox, 15, 18
Rosedale, Philip, 58–59
Russia, 106–107

Safe Zone, 60–61
scams, 56–58
scarcity, 10–14
Schneier, Bruce, 48, 62
science fiction, 136–138
search costs, 93
Searls, Doc, 140
security. *See* cybersecurity
self, physical and digital, 68–69
self-expression. *See* speech and expression
self-governance, 111–114
sexual harassment, 60–61
Snow Crash (Stephenson), 2, 138
social engineering, 56–58
software/hardware complementarity, 16
spatial audio, 32, 39–40
spatial data, 94
speech and expression, 31
   and AI bots, 133–135

content moderation, 38–41
   legal impacts and modernizations, 41–47
   opportunities and challenges, 32–38
stalking, 44–45
Stephenson, Neal, 2
stereotypes, biases, and discrimination, 73–74
substitutes, 28–30
supply and demand, 14–18
Sweeney, Tim, 6

targeted ads, 92–94
taste heterogeneity, 22
technology inconsistencies, 19–20
therapy, 34, 94–95
third parties, data-selling to, 84–85
three "Bes," 65
tort law and criminal law, 44–45
touch, 39
trademark concerns, 42–43, 61–62
tragedy of the commons, 112
2030 Digital Compass, 117

United Metaverse Nation, 3–4
usage data, 82

Velasquez, Eva, 54–55
virtual human services, 16–17
virtual reality (VR) therapy, 34, 94–95
virtual spaces/environments, 36–37
virtual-world games, 127–129
Vision Pro headset, 9, 21, 126–127
VPNs (virtual private networks), 59, 62
VR (virtual reality)
   headsets. *See* headsets
VR (virtual reality) therapy, 34, 94–95

Web 2.0 and 3.0, 107–108
World Summit on the Information Society (2005), 107

Yuanverse, 1–2

Zelnick, Strauss, 6–7
zero trust security, 50–51

Zuboff, Shoshana, 92
Zuckerberg, Mark, 5, 7–9, 104–105, 124–126